MW01527868

MEDICAL LIBRARY SERVICE

COLLEGE OF PHYSICIANS AND SURGEONS
OF BRITISH COLUMBIA

Handbook of Toxicology and Ecotoxicology for the Pulp and Paper Industry

Handbook of Toxicology and Ecotoxicology for the Pulp and Paper Industry

by

LAURA ROBINSON
Toxicologist, Eka Chemicals AB, Sweden

and

IAN THORN
General Manager, New Markets, Eka Chemicals, UK

Medical Library Service
College of Physicians & Surgeons of B.C.
1383 W. 8th Ave.
Vancouver, B.C.
V6H 4C4

b

Blackwell
Science

© 2001 by
Blackwell Science Ltd
Editorial Offices:
Osney Mead, Oxford OX2 0EL
25 John Street, London WC1N 2BS
23 Ainslie Place, Edinburgh EH3 6AJ
350 Main Street, Malden
  MA 02148 5018, USA
54 University Street, Carlton
  Victoria 3053, Australia
10, rue Casimir Delavigne
  75006 Paris, France

Other Editorial Offices:

Blackwell Wissenschafts-Verlag GmbH
Kurfürstendamm 57
10707 Berlin, Germany

Blackwell Science KK
MG Kodenmacho Building
7–10 Kodenmacho Nihombashi
Chuo-ku, Tokyo 104, Japan

Iowa State University Press
A Blackwell Science Company
2121 S. State Avenue
Ames, Iowa 50014-8300, USA

The right of the Authors to be identified as the
Authors of this Work has been asserted in
accordance with the Copyright, Designs and
Patents Act 1988.

All rights reserved. No part of
this publication may be reproduced,
stored in a retrieval system, or
transmitted, in any form or by any
means, electronic, mechanical,
photocopying, recording or otherwise,
except as permitted by the UK
Copyright, Designs and Patents Act
1988, without the prior permission
of the publisher.

First published 2001

Set in 10/12pt Times
by DP Photosetting, Aylesbury, Bucks
Printed and bound in Great Britain by
MPG Books, Bodmin, Cornwall

The Blackwell Science logo is a
trade mark of Blackwell Science Ltd,
registered at the United Kingdom
Trade Marks Registry

DISTRIBUTORS

Marston Book Services Ltd
PO Box 269
Abingdon
Oxon OX14 4YN
(*Orders:* Tel: 01235 465500
         Fax: 01235 465555)

USA
Blackwell Science, Inc.
Commerce Place
350 Main Street
Malden, MA 02148 5018
(*Orders:* Tel: 800 759 6102
         781 388 8250
         Fax: 781 388 8255)

Canada
Login Brothers Book Company
324 Saulteaux Crescent
Winnipeg, Manitoba R3J 3T2
(*Orders:* Tel: 204 837-2987
         Fax: 204 837-3116)

Australia
Blackwell Science Pty Ltd
54 University Street
Carlton, Victoria 3053
(*Orders:* Tel: 03 9347 0300
         Fax: 03 9347 5001)

A catalogue record for this title
is available from the British Library

ISBN 0-632-05436-0

Library of Congress
Cataloging-in-Publication Data
Robinson, Laura.
    Handbook of toxicology and ecotoxicology
  for the pulp and paper industry/by Laura
  Robinson and Ian Thorn.
      p. cm.
    Includes bibliographical references and index.
    ISBN 0-632-05436-0
    1. Toxicology—Handbooks, manuals, etc.
  2. Environmental toxicology—Handbooks,
  manuals, etc.   3. Paper industry—Health
  aspects—Handbooks, manuals, etc.
  4. Pulpwood industry—Health aspects—
  Handbooks, manuals, etc.   I. Thorn, Ian.
  II. Title.
    RA1215 .R63 2001
                                        00-052932

For further information on
Blackwell Science, visit our website:
www.blackwell-science.com

Contents

Preface

The aim of this book is to provide a guide to both the interpretation and application of toxicological and ecotoxicological information provided by suppliers of chemicals and to improve the understanding, and therefore the awareness, of the possible implications that handling a given chemical has on the environment and human health. The difference between test protocols and the criteria used to determine the intrinsic hazards of chemicals to both health and the environment is highlighted.

There is a growing awareness in the world at large (which is reflected within the pulp and paper industry) of the possibility that side effects can be experienced from the use of chemicals. The controversy highlighted in the media surrounding the use of gene-modified foods is the most recent manifestation of this. The pulp and paper industry too, has not escaped the opprobrium of environmental insensitivity that has accompanied the efflux of dioxins from bleached pulp mills into surrounding rivers and lakes.

All responsible suppliers of chemicals to the industry strive to maintain the information flow necessary for the safe handling and end-use of the paper product. For example, those purchasing chemicals are often provided with a Safety Data Sheet from which information on the potential adverse effects to human health and to the environment can be obtained. However, as products are often manufactured in different parts of the world, the information provided by the test results in the Safety Data Sheet may be based on different guidelines to those required by the regulations current in the country in receipt of the chemicals. This can give rise to different terminologies which sometimes hinder the interpretation of the latent hazards.

This book provides the paper technologist, health and safety manager or technical manager with as comprehensive a book as possible of all the chemicals in use in the industry, together with their properties – given the twin constraints of new chemicals being introduced all the time whilst others are superseded and the degree of secrecy which exists within the industry.

In conclusion, we hope that our efforts will contribute to the safety of both workers in the industry and the environment surrounding it.

Acknowledgements
We would like to thank BASF, Clariant, Hercules, Buckman, EKA Chemicals and those other chemical suppliers who sent us their Material Safety Data Sheets. We would also like to thank Dr Colin Rogers for kindly contributing his chapter on Regulatory affairs, Professor Lars Renberg, Dr Thomas Vierling and colleagues within EKA Chemicals and Akzo Nobel for their help, Todd Levy for helping out with some of the illustrations and last, but not least, Dr Shirley Price of the University of Surrey UK, Department of Biological Sciences, for her invaluable assistance. Dr

Price is the Course Organiser for the Modular Training Programme in Applied Toxicology, which is of outstanding value for those who wish to gain more knowledge in this area.

Laura Robinson
Ian Thorn

For our parents,
Todd and Sally

1 Toxicology

1.1 Introduction

Toxicology can be defined as the study of poisons, which to many probably conjures up images of witches hovering around a cauldron, cooking up a lethal potion for some unsuspecting victim!

Therefore, perhaps a better and more appropriate definition for the purpose of this book is:

Toxicology is the study of adverse effects of chemicals on living organisms.

Owing to the strict legislation that operates in most countries, chemical companies wishing to introduce their products onto the market have to assess firstly the hazards associated with them. (In this context, 'hazards' means adverse health effects.)

Toxicity can be defined as the 'inherent capacity of a chemical to cause adverse health effects'. Therefore, with relevance to humans, toxicologists are employed to investigate the potential adverse effects and associated mechanisms of toxicity arising from exposure to chemicals.

For toxicity to occur there has to be exposure to the chemical and there are three main ways in which this can occur: inhalation, skin contact and ingestion. Some chemicals cause toxicity directly whereas others require conversion to a more reactive form. Once *exposure* has taken place a chemical can cause *local effects* such as irritation at the site of contact, otherwise it may be *absorbed* into the bloodstream and distributed to other parts of the body which could result in *systemic effects* (see Table 1.1). Note that it is not uncommon for a chemical to cause both local and systemic effects.

Table 1.1   Two common types of adverse effects caused by chemicals.

Local effects	These occur at the point of exposure. Examples include irritation and corrosivity.
Systemic effects	Effects that occur in cells, tissues or organs. Effects occurring in specific organs are called 'Target Organ Effects'. For instance, carbon tetrachloride (tetrachloromethane) is a hepatotoxin (causes toxicity in the liver), although a chemical may affect more than one target organ.

Not all chemicals can be absorbed, and there can be large differences in the extent and rate at which absorption occurs. As an example, there is nearly no absorption whatsoever for most polymers, whereas for other chemical substances such as alcohol, full absorption occurs.

Once inside the body a chemical may be *metabolised*, stored or else *excreted*. Metabolism involves enzymes that convert the chemical into a form (called a metabolite) which is more readily excreted. However, the problem with metabolism is that it can sometimes convert the chemical into a more reactive toxic form. Some chemicals can also be stored within the body, which can in some cases cause cumulative toxicity. Excretion (in the breath, urine or faeces) leads to the removal of the chemical or its metabolites from the body.

1.1.1   *Toxicity testing*

In order to protect the general population, many countries have strict requirements for the hazardous nature of chemicals to be established prior to their use. The aim of toxicological testing is to predict the toxicity in man by using animals or cell cultures as surrogates. The tests themselves aim to mimic as closely as possible the likely exposure scenario that a human being would face, using the most appropriate animal model (or in some cases, cell cultures). There are standard test procedures, which can be used for toxicological evaluations and some of these procedures use animals, whilst others use cultured cells and other *in vitro* techniques Throughout this book, the Organisation for Economic Co-operation and Development (OECD) test guidelines will be referred to as these are accepted in most countries.

Different species of animals are used in toxicity testing because there is no one animal that is satisfactory for use in the evaluation of all toxicological effects. The choice will depend on the type of toxicity to be investigated, although other factors such as availability, cost and reliability in their response will also influence the choice. Both sexes are usually employed although in some cases tests may be gender-specific. In tests where more than one dose level is used, they are selected so as to establish a *dose–response relationship* and *threshold level* of effect. Note that these terms will be described more fully in subsequent sections.

The aim of the following sections is to describe some of the different toxicological effects that can arise as a result of chemical exposure. This will, hopefully, enable the reader to better understand and use the toxicological information provided by chemical suppliers.

1.2   Toxicity

Many people perceive all chemicals as being bad in some way or another. The expression 'toxic chemicals' is often used in everyday life, and questions concerning whether or not one particular chemical can cause more harm than another are frequently raised. The kinds of questions raised were studied as far back as the sixteenth century by a Swiss scientist named Paracelsus and the results of his work are now considered to be an underlying theme in the field of toxicology. He concluded that in sufficient quantities everything has the potential to cause adverse health effects and that the only thing that differentiates something from being harmful or not is the dose. In other words, 'it is the dose which makes the poison'.

This may seem like a strange concept as many of us have the idea that when dealing with a 'toxic chemical', its mere presence in our vicinity can cause harm to us. But this is not the case and, as was written by Paracelsus, it depends on the exposure dose as there will be a dose level below which this 'toxic chemical' will not cause us harm. (There are exceptions, however – genotoxic carcinogens are believed to be active at all dose levels.) For some chemicals such as arsenic, which features in so many murder/suspense movies, this dose level is very low, whereas for other chemicals such as sugar, the dose level is much higher because it is less toxic. Conversely, so-called 'harmless' chemicals, which are around us all the time, such as water, oxygen and even those which we may daily use in the kitchen such as table salt, have the potential to cause us harm should the amount to which we are exposed be sufficiently high.

1.2.1   *Acute and chronic effects*

Toxicity can be defined as the propensity to cause harm (or adverse health effects). There are two main types of toxicity that need to be taken into consideration when studying the adverse effects of chemicals. These are acute toxicity and chronic toxicity.

- *Acute toxicity* describes the adverse health effects following a single or limited number of exposures.
- *Chronic toxicity* describes the adverse health effects resulting from continuous or intermittent exposure over a lifetime. An example of chronic toxicity is organ damage, such as liver cirrhosis arising from long-term alcohol abuse.

Both acute and chronic toxicities are very different with respect to how they occur, i.e. their manifestation, the target organs involved and also the resulting adverse effects.

For example, acute effects tend to appear quickly and can be reversible, whereas chronic effects usually take a longer time to appear and are often irreversible. Therefore, it is not possible to predict the chronic toxicity of a particular compound based on its acute toxicity, or vice versa – which is a question that is often asked! In fact, both these forms of toxicity can be considered to be extremes of each other, the differences being based on the dose levels experienced and the exposure period. Between these two types of toxicity lie two other types of toxicity, and two other terms, which are often seen in relevant literature. These are *subacute toxicity* and *subchronic toxicity*.

Subacute and *subchronic* toxicity describes the adverse health effects arising from daily or frequent exposures (to smaller levels of chemical relative to acute toxicity) over part of a lifetime. The difference between these two toxicities lies in the duration of exposure.

1.2.2   *What factors influence toxicity?*

Every day we are exposed to a wide variety of chemicals such as detergents, hand cream, shampoo, chemicals in the food we eat and the air we breathe, etc. However, what should be noted is that chemical exposure does not always give rise to an adverse health effect. There are a number of factors that will influence the outcome of chemical exposure and whether or not a toxic effect will occur. These are considered below.

1.2.2.1   The dose   The dose or amount of chemical to which an individual is exposed is the one factor that has the greatest influence on toxicity, and this had already been discovered by Paracelsus way back in the sixteenth century. What is usually seen experimentally is that at lower chemical dose (or exposure) levels there are no toxic effects. However, as soon as the dose or exposure increases, so does the possibility of the occurrence and severity of a toxic response. This can be expressed graphically in the form of a dose–response curve and this is a major concept in toxicology and forms the basis for all toxicity tests.

The OECD guidelines define dose–response as 'the relationship between dose and the proportion of a population sample showing a defined effect'. Experimentally, the magnitude of the effect and/or the number of test animals affected will increase with increasing dose as can be seen in Fig. 1.1 which shows the effect of increasing dose against the percentage response in terms of mortality. Theoretically, a classic sigmoid-shaped curve is obtained, although in practice this is not often seen.

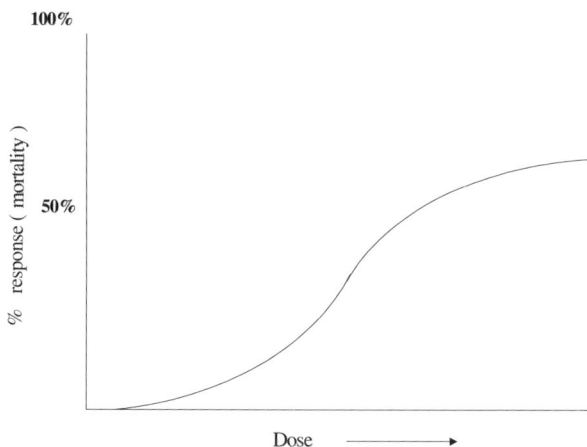

Fig. 1.1   A typical dose–response curve where, in this case, the response is mortality.

1.2.2.2 *Chemical structure* This will also be a significant factor because the chemical structure will dictate how the chemical reacts with the body (i.e. toxic effects) and how it is both metabolised and excreted (if at all). In the paper-making process polyacrylamides are often used as a retention aid. The polymer itself is a relatively toxicologically inert structure, made up of acrylamide monomers. However, the monomers are highly reactive (all monomers are by their very nature). In Europe the acrylamide monomer is classified as 'Toxic', whereas the polymer is not considered to be hazardous to health. Chapter 6 provides both toxicological and ecotoxicological information for both a typical polyacrylamide and an acrylamide monomer. The difference in the toxicological effects speaks for itself.

1.2.2.3 *Route of exposure* The most common routes of exposure are inhalation, ingestion (to a lesser extent in the industrial setting) and skin contact. For each exposure route there are usually different target organs involved and fortunately only very few chemicals are toxic by all routes of exposure.

1.2.2.4 *Host factors* Very young or old people are more susceptible to toxic chemicals, one principal reason being that their metabolism and excretion function is not very efficient, which means that there is a risk of toxic chemicals building up in the body leading to cumulative toxic effects. Any predisposing illnesses, which may affect metabolism and/or excretion, will also influence the outcome. Likewise, those who are poorly nourished are more susceptible to the toxic effects of chemicals, again because they are less effective in metabolising and excreting undesirable compounds.

1.2.3 *Testing for these different types of toxicity and the information obtained*

When it comes to conventional toxicity testing, the types of toxicity are often categorised into three groups.

(1) Acute studies – for acute toxicity.
(2) Short-term (repeated dose) studies – for subacute or subchronic toxicity.
(3) Long-term studies – for chronic toxicity.

1.2.4 *Which exposure route will be chosen?*

As with any toxicological test, we ideally want to mimic the likely route of exposure when handling or using chemicals in the occupational setting (excluding other areas where chemicals are used). When it comes to handling chemicals in the workplace, it probably comes as no surprise to learn that the most likely routes of exposure are either by skin contact or by inhalation. Ingestion is itself not usually considered to be a relevant route of exposure in the workplace, although of course, it can never be totally ruled out, as there will always be incidences of accidental ingestion.

Any literature survey will show that it is the oral route that is the most common route of exposure used in these studies. This is partly because dosing by the oral route is inherently easier to carry out, and is also cheaper when compared to the other test routes. Also, in the past, most interest was focused on the use of chemicals in the food industry and of course the logical dosing regime involved the oral route. It has only really been in recent years that the importance of exposure by both inhalation and by skin contact, especially in the workplace, has come to light and that consequently, these routes have also been taken into consideration when carrying out toxicity tests. However, the oral route is still used as a 'standard' when attempting to compare relative toxicities of different chemicals and this is what regulators will consider as one of their requirements.

Acute inhalation studies are not at all common because they can only be performed on certain types of chemicals and they are also very expensive to carry out because specialised staff are required both to do the tests and to analyse the results.

1.2.5 *Acute studies*

1.2.5.1 *Acute toxicity testing*   The purpose of acute toxicity tests is the same, regardless of the chosen route of exposure. That is, they are undertaken to investigate the potential adverse effects arising from exposure to a given chemical over a short period of time. There are many different types of acute effects that could be studied, but the one acute effect, or end-point, which all chemicals will demonstrate is lethality and it is this which is therefore used as the end-point in this type of study. The results from acute toxicity tests are graphically represented as a dose–response curve which is often converted to a straight-line plot as it makes the data easier to handle and interpret.

1.2.5.2 *The LD_{50} test*   This is probably one of the most well-known toxicological tests that you will come across. The term LD_{50} (lethal dose, 50) is used to describe the acute oral or dermal median lethal dose, i.e. the single lethal dose which will kill 50% of the test population. Its value is usually expressed in milligram or gram of test compound per kilogram of animal weight ($mg \ kg^{-1}$). However, for acute inhalation studies the value used is LC_{50}, which refers to the median lethal concentration in air. It is usually expressed either as parts per million (ppm) or $mg \ m^{-3}$ (milligrams of chemical per cubic metre of air). Both the LD_{50} and LC_{50} values are determined from the dose–response graph as mentioned earlier.

1.2.5.3 *How is this test carried out?*   The tests involve the use of three groups of test animals (10 animals per group, usually rodents) which are administered increasing graduated doses of the test chemical, one per dose group. After an observation period of 14 days, where all mortalities are noted along with any behavioural effects, etc., all the animals are autopsied and a percentage response (lethality) against dosage administered is plotted. From this the LD_{50} value is derived as is shown in Fig. 1.2.

The OECD test guidelines provide the full details for these tests and they are quite 'reader-friendly'. It should be noted that once the dose–response curve has been established, other values can be derived, such as the LD_0, the dose at which no deaths occur, etc.

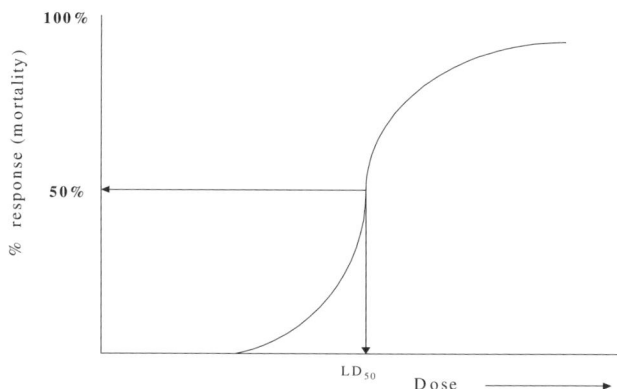

Fig. 1.2   Derivation of an LD_{50} value from a dose–response curve.

1.2.5.4   *What information do we get from an acute toxicity test?*   An acute toxicity (LD_{50}) test provides information on the magnitude of the acute toxic dose, i.e. the LD_{50}, which can then be used for classification and labelling purposes and can also be used as a rough measure of relative toxicity. The results from these types of studies can also provide some insight into any systemic toxicity arising from that particular exposure route, and can be used in order to establish the dosing strategy for other repeat-dose toxicological studies.

In the real world, i.e. in terms of human exposure, these results can be used in a risk assessment in order to help predict what might happen as a result of an accident involving contact with or inhalation of large amounts of chemical or accidental (or otherwise) ingestion. The LD_{50} value obtained in such tests will depend on the choice of test species (because some are more sensitive than others) and also on the chosen route of exposure. For example, some chemicals will exhibit toxicity by one exposure route but not by another. An example of such a chemical is vitamin D. This vitamin is not acutely toxic by the dermal route but it is acutely toxic by ingestion.

1.2.5.5   *How to interpret the results*   Interpreting the LD_{50} (or$_{50}$ LC_{50} for acute inhalation) value is quite easy to do. Simply put, the lower the LD_{50} value the more toxic the test compound and the higher the LD_{50} value, the less toxic the test compound.

Therefore, a chemical with an LD_{50} value of $5\,mg\,kg^{-1}$ is more toxic than one with an LD_{50} value of $300\,mg\,kg^{-1}$ because fewer milligrams of chemical per kilogram of body-weight are needed to cause death (for that particular exposure route).

1.2.5.6   *The limit test and fixed dosing method*   Over the years there has been a lot of criticism of the acute LD_{50} test. This has mainly been from an ethical standpoint in that a large number of animals are used and the end-point is lethality in order to obtain a numerical value (the LD_{50}), a value which is only relevant for those particular test conditions.

The limit test essentially aims to reduce the numbers of animals involved in testing, although lethality is still the end-point for those that are used. This is accomplished by the use of smaller groups of animals using dosage levels up to a pre-set limit (2000 mg kg^{-1} for acute oral tests). The fixed dosing method (for acute oral toxicity) was introduced in 1985 by the British Toxicological Society in a bid to address the ethical issues and validity of the LD$_{50}$ test. It has since then been adopted into an OECD procedure, OECD 420.

This test uses a series of predetermined dosing levels and relies on obvious signs of toxicity instead of lethality. Therefore, there is no LD$_{50}$ derived from this study. This fixed dosing procedure is likely to replace the more traditional LD$_{50}$ acute oral study in the future and is therefore worth knowing about now.

1.2.6   Short-term (repeated dose) studies

Subacute or subchronic studies are designed to investigate the adverse effects resulting from repeated exposures to smaller levels of chemical (relative to acute toxicity studies) over part of the organism's lifetime (usually under 10% of the life-span). These tests aim to mimic the exposure pattern of humans who may daily work with and are exposed to low levels of chemicals, which in reality is probably more representative of the actual daily workplace exposure scenarios that can occur than is the case with acute toxicity studies.

1.2.6.1   NOELs and LOELs – what are these?   One feature that both short-term studies and long-term studies have in common is that in contrast to acute toxicity studies, these tests do not use lethality as the end-point. Therefore, no 'LD$_{50}$ type of value' is generated. Instead, what these studies attempt to do is to find the smallest, or lowest dose which produces any kind of detectable adverse effect, whether behavioural, some kind of organ damage or a change in weight, etc. This value is known as the *lowest observable effect level* (LOEL). As well as this LOEL value, it is also important to know the highest dosage level at which there will be no detectable adverse effects. This is known as the *no observable effect level* (NOEL).

1.2.6.2   How are short-term tests carried out?   Groups of test animals (usually rodents) are daily administered graduated doses of the test chemical over a period usually amounting to one-tenth of their total lifetime. The doses administered to each group are varied so that a dose–response curve can be constructed. From this it is possible to determine the dose at which the first detectable toxic response occurs, which is of course the LOEL, and the highest dose where there is no toxic response, the NOEL. The objective of both subacute and subchronic studies is the same, but the test duration and animal numbers used differ. According to OECD guidelines and assuming that the rat is used as the test animal and dosing is by the oral route, subacute studies usually last for a period of 28 or 14 days, using 10 rodents per group (OECD 407), while subchronic studies last 90 days and use eight test animals per group (OECD 408).

1.2.6.3 *What information do we get from short-term studies?* These short-term tests enable the determination of both the LOEL and the NOEL. They can also help pinpoint any potential target-organ effects, i.e. those organs where the chemical produces its adverse effects, and help determine whether there is any possible accumulation effects. This is important because it is possible for a chemical to be of very low acute toxicity (high LD_{50} value) but as a result of accumulative effects it can interfere with critical bodily functions thereby producing some form of 'delayed toxicity' which would not have been detected by a short-term study. Of course, the results from such tests can be used in selecting the appropriate dose levels for longer-term studies.

1.2.7 *Long-term studies*

Chronic studies investigate the adverse effects arising from prolonged or repeated exposure to low levels of chemical over the whole or the greater part of a lifetime. These studies allow time for any adverse effects, which have a long latent period, to show up along with any cumulative effects.

As can be imagined, results from chronic toxicity studies will be of special interest to us as consumers when talking about food additives, pesticide residues on fruit and vegetables, etc. to which we are all daily exposed throughout our lives. The studies themselves are similar to the subacute/subchronic tests in terms of the design protocol, the major differences being of course the test duration, dosage levels and the number of test animals used. The reader is referred once more to the OECD guidelines for more information. Any literature survey will show that it is the oral route that is the most common route of exposure used in these studies. As can probably be recalled, this was also the case in acute toxicity studies and the reasons are the same.

Since chronic studies are extremely expensive to perform and involve the use of many test animals, it is actually not uncommon that they are combined with carcinogenicity studies. In this way the design of the test is such that the objectives of the two separate studies are completely fulfilled.

1.2.7.1 *What information is obtained from long-term studies?* Similar to the repeated dosing studies, chronic studies provide information on adverse effects resulting from prolonged or repeated exposures and also on any effects that have a long latent period or are cumulative. These tests can also provide information on dose–response relationships and can in turn be used for the determination of LOEL and NOEL values.

1.2.7.2 *Chronic toxicity and carcinogenicity* Because of public concern over cancer, and especially any risk of developing it as a result of chemical exposure, carcinogenicity is treated separately from other chronic effects. The tests used are also very different because they aim to detect the probability of one person in one million developing cancer as a result of exposure to a suspected carcinogen. As can be imagined, in order to achieve this kind of sensitivity in a test, a huge number of animals would be needed which could never be acceptable ethically, economically or even from a practical perspective. Therefore, these carcinogenicity tests utilise very high doses as a means of overcoming the relatively small numbers of animals used and ensuring that the results are statistically relevant (refer to Section 1.5.4).

Unlike other chronic toxicity studies, carcinogenicity studies do not aim to detect safe levels of exposure. Instead, they aim to see whether or not there is an increased incidence of any type of cancer as a result of chemical exposure.

1.2.7.3 *Is it possible to predict the chronic toxicity from an acute or subacute study?* No, it is not possible to predict the potential adverse health effects arising from chronic exposure to a chemical based on acute toxicity test results.

Different effects can arise when exposure to a chemical is repeated over a period of time as compared with a single exposure to a large dose of chemical, which is characteristic of an acute toxic exposure, and therefore the mechanisms involved can be quite different. Also, since chronic toxicity studies take place over the whole lifetime or for the greater part of the test animal's life, the ageing process itself, as well as any spontaneous diseases which are related to it, may have an influence on the final outcome, which would not be detected in an acute toxicity study as the test duration is so much shorter.

Further reading

Occupational Toxicology
Edited by Neill H. Stacey
Taylor & Francis, 1993

Basic Toxicology: Fundamentals, Target Organs and Risk Assessment, 3rd edn
Frank C. Lu
Taylor & Francis, 1996

The Dose makes the Poison: A Plain Language Guide to Toxicology
M. Alice Ottoboni
Van Nostrand Reinhold, 1991

Mechanisms and Concepts in Toxicology
W. Norman Aldridge
Taylor & Francis, 1996

General and Applied Toxicology, abridged edn
Edited by Bryan Ballantyne, Timothy Marrs & Paul Turner
Macmillan Press, 1995

OECD Guideline for Testing of Chemicals
Acute Oral Toxicity, No. 401

OECD Guideline for Testing of Chemicals
Acute Dermal Toxicity, No. 402

OECD Guideline for Testing of Chemicals
Acute Inhalation Toxicity, No. 403

OECD Guideline for Testing of Chemicals
Repeated Dose Oral Toxicity – Rodent 28-day or 14-day, No. 407

OECD Guideline for Testing of Chemicals
Subchronic Oral Toxicity, No. 409

OECD Guideline for Testing of Chemicals
Chronic Toxicity Studies, No. 452

OECD Guideline for Testing of Chemicals
Combined Chronic Toxicity/Carcinogenicity Studies, No. 453

1.3 Chemical irritancy and corrosive effects

1.3.1 *The skin*

The skin is one of the largest organs of the body. It provides both an external barrier to pathogens and chemicals with which we come into daily contact and also an internal barrier, permitting only the 'escape' of water and electrolytes, in the form of perspiration. It is not, as many think, a single thin layer which 'keeps everything intact'. Instead, it is multilayered and carries out some vital functions, including the synthesis of vitamin D, protection, body temperature regulation, etc. Because of this, any adverse effects (or toxic changes) in the skin will be of vital importance to humans.

As can be seen from Fig. 1.3 the epidermis and the dermis rest over the subcutaneous layer. The *epidermis* comprises an upper layer of dead, cornified cells, which is known as the stratum corneum and a lower, living (viable) layer. The *stratum corneum* is the main protective layer of the skin and acts as a barrier to the passage of most chemicals, electrolytes and water. It typically has a water content of between 10 and 20% in ambient conditions, but this can increase greatly, such as during high humidity, which can lead to a greater permeability.

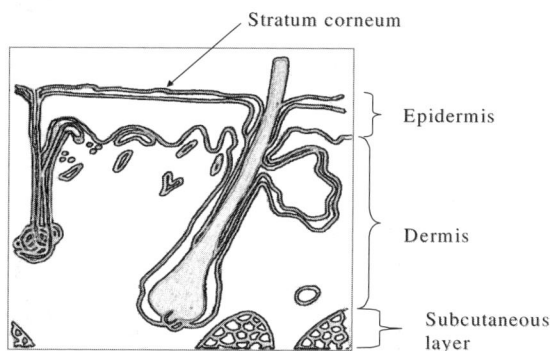

Fig. 1.3 The main layers of the skin.

Its thickness can vary from species to species and also in different areas of the body. As an example, the stratum corneum is much thicker on the palms of our hands than it is on our foreheads.

The epidermis also contains other cells. These include dendritic cells (known as Langerhans cells), which are an important part of the immune system, and the melanocytes, which produce pigments. The *dermis* forms the main part of the skin. Unlike the epidermis, it is rich in blood vessels, and comprises lymphatic and fibrous (collagen) material, which provides support. There are different cell types present within the dermal layers. These include macrophages, lymphocytes and mast cells, which will respond to chemical injury.

Intact skin is an effective barrier against many chemicals. However, even when intact, it is not a complete barrier and some chemicals, such as acetone, are able to penetrate. Similarly, any pre-existing skin condition, which involves the epidermis or minor abrasions of the skin, will considerably enhance the absorption of many chemicals.

1.3.1.1 *How chemicals cause adverse effects when in contact with the skin*   There are two possible ways in which chemicals that come into contact with skin can produce adverse health effects. The first is by their absorption (assuming this occurs of course), which can lead to systemic effects. In this case, the extent of absorption will be a major factor in determining the degree of toxicity. It should also be noted that in some cases, chemicals can exert their effects via the sebaceous glands, sweat glands and even the hair follicles.

Secondly, chemicals may react with the skin itself, giving rise to a localised inflammatory reaction, called *contact dermatitis*. This may be caused either by an allergic reaction (*allergic contact dermatitis*) or by irritation (*irritant contact dermatitis*). Typical signs of contact dermatitis include redness of the skin, blisters, rashes and crusts, along with a burning or itching sensation. Not all of these symptoms necessarily occur at the same time or in all cases.

1.3.1.2 *Cutaneous irritancy*   Irritants can be defined as 'those substances which, when applied to normal skin for long enough (either continuously or discontinuously) and in high enough concentration, will cause, in all subjects, a non-immunological inflammatory reaction' (Hunter & Aldridge, no date).

Probably the most common chemical 'irritant' that comes to mind for most people is an acid or alkali. For these, the concentration used will determine the severity of the effects. An acid such as hydrochloric acid with a 0.1 M concentration will cause redness and sting when it comes into contact with the skin. A higher concentration will produce corrosive effects (corrosivity is described later). However, it is not just these chemicals that have been classified as an 'irritant' according to set criteria (refer to Chapter 3). Under certain circumstances many different kinds of chemicals such as organic solvents, surfactants, abrasives and desiccants can act as irritants, even though they may not be classified as such.

Some chemical substances will cause a local inflammatory reaction as a result of a single contact with the skin in all those individuals who are exposed. Contact with these kinds of irritant are unlikely to go unnoticed because it would almost certainly be a painful experience. Such chemicals are called *acute primary irritants*. Sulphuric acid and sodium hydroxide are examples of acute primary irritants.

Other chemical substances require multiple or prolonged skin contact before dermatitis occurs. This is known as *cumulative irritation* (or cumulative irritant dermatitis). Unlike acute primary irritation, this could go undetected for some considerable time even though slow adverse changes in the skin would be taking place as a result of the prolonged or repeated contact. Single exposure to these kinds of chemicals would not be sufficient to produce dermatitis. *Marginal irritant* is the name given to the type of substance producing this kind of dermatitis. Examples include detergents and soaps. However, even physical factors such as humidity, friction and cold can cause cumulative irritant dermatitis. Those individuals who have an atopic background are also susceptible to cumulative irritation dermatitis.

1.3.1.3 *What can promote the development of irritant contact dermatitis?* There are certain intrinsic and extrinsic factors that contribute to the development of irritant contact dermatitis.

Intrinsic factors are related to the individual who is exposed. As has already been mentioned, some individuals are more prone to developing cumulative irritant dermatitis than others. These individuals tend to have some atopic history (allergies). Such individuals could develop dermatitis as a result of repeated contact with detergents or soaps, whereas non-atopic individuals would not experience any problems. This kind of reaction is not at all uncommon. Other factors specific to the individual include genetic background, age and region of the skin exposed. The health of the skin is also of importance. This is because any pre-existing skin disorders could promote the development of a dermatological reaction.

The extrinsic factors are related to the nature of the chemical substance to which exposure occurs and also to the working environment itself. The concentration of the chemical, pH, solubility and circumstances of contact (occlusion, length and frequency of exposure) will be of importance. Another factor, which should be taken into consideration, is the degree of absorption.

Some chemical substances have the ability to remove both moisture and oils from the outer skin layers, thereby reducing the protective barrier capacity. The result is an increase in the absorption of chemicals into the lower layers of the skin.

The working environment itself can also play an important role. As an example of this, consider the humidity of the working environment. A hot, humid environment will increase the level of sweating and this may either increase or decrease the irritant effects of chemicals that come into contact with the skin. This is because, depending on the circumstances, the sweat can either dissolve the chemical or promote its absorption, or it can protect the skin by simply washing it away. Low air humidity will also enhance the possibility of irritation, but this time as a result of skin chapping.

In fact, any working situation in general where skin chapping may occur, such as during instances where the skin is repeatedly wet over prolonged periods of time, can increase the likelihood of irritation and the development of cumulative irritant dermatitis. Thus, even water can occasionally be implicated in the development of this condition! This is often seen in professions involving so-called 'manual wet work', such as hairdressing or cleaning, etc.

Another aspect to take into consideration is *occlusion*. As soon as you put on a pair of protective gloves, you occlude the skin and in turn increase the hydration of the stratum corneum, which is the main protective outer skin layer. This is because the skin of the hand becomes warmer and more moist which increases its permeability to chemicals. In fact, it can significantly increase the permeation of chemicals. Therefore, it is essential that the hands are thoroughly clean, i.e. they are free of any prior chemical contamination, and completely dry before gloves are used.

1.3.1.4   *A case in point – highlighting the effects of occlusion*   An incident in a paper mill that involved a technical sales representative who had been handling a chemical which is a skin sensitiser can be cited as an example. The pump used with this particular chemical had become blocked. Forgetting to put on protective gloves, the representative unblocked the pump. During this a small amount of chemical came into contact with the representative's hands, whereupon gloves were put on. One day later the hands of the representative had broken out in a rash and the skin had begun to peel. The effects of the chemical had been enhanced owing to the occluded effect provided by the gloves. So, the lesson to be learnt here is that should you initially forget to wear the appropriate gloves and wish to start using them, make sure that your hands are completely free from chemical contamination before putting them on!

1.3.1.5   *Corrosivity and irritancy – what is the difference?*   One question which is often asked: 'What is the difference, assuming the same exposure conditions, between a chemical which is irritating and one which is corrosive?'

In simple terms, a chemical which is an acute primary irritant produces local, reversible inflammatory effects. A corrosive chemical, however, visibly and irreversibly destroys all living tissues with which it comes into contact (corrosive effects are often referred to as 'chemical burns').

Regardless of whether the skin, eye or respiratory tract is exposed to a corrosive chemical, the above definition still holds true.

1.3.1.6   *Tests for skin irritancy and corrosion*   Take just five seconds and think about all the chemicals with which we come into contact on a daily basis. Soaps, detergents, cosmetics, and solvents, all of which have some potential to affect the body in some way (and not just to make it smell better!). These effects can either occur locally, at the site of contact, or elsewhere in the body. Therefore, it is vitally important to have information regarding their ability to cause harm during the course of customary use or even anticipated misuse.

Probably one of the most well-known animal tests for predicting potential skin irritants is the Draize test, details of which were published in 1944. The OECD 404 test procedure, which is based on this, involves the administering of the test chemical either directly onto a small area of shaved skin or onto gauze patches that cover the shaved area. One or more test animals are used, usually the albino rabbit, and the duration of exposure is four hours under semi-occluded conditions, whereupon any remaining traces are removed. The animals are then observed for any signs of irritation and resolution at specific time intervals for up to 72 hours after the exposure. Grading of any skin response is based on erythema/eschar and edema formation and a grading scheme is provided in the test method.

It is very common to see in Material Safety Data Sheets (MSDS) that a chemical is referred to as 'mildly irritating' or is a 'severe irritant'. This comes from the Primary Irritation Index as shown in Table 1.2. This is the combined average scores for total erythema and edema formation that resulted at the time points of 24 and 72 hours.

Table 1.2 The Primary Irritation Index (Draize, 1959).

Index value	Effect
2	Mildly irritating
2–5	Moderately irritating
6	Severe irritant

For ethical reasons, if the test chemical is known to be, or is suspected of being corrosive, for example by its pH (if ≤ 2.5 or ≥ 11.5) then this test would not be required. One criticism of this type of test is that the albino rabbit is more sensitive than other species to irritants, which suggests that a chemical deemed to be a mild or moderate irritant to the rabbit may produce no such reaction when applied to humans. It should also be noted that this test will only detect primary irritants. It will not detect those chemical substances that are marginal irritants.

1.3.2   The eye

The eye has one main function – photosensory reception. It is a complex structure made up of three main layers (Fig. 1.4). The outer layer comprises the sclera and the cornea, which are transparent. The middle vascularised layer comprises the iris, the choroid and ciliary body. The retina makes up the inner sensory layer. Chemicals can affect the eye either locally or systemically.

Of particular importance to those of us who handle chemicals in the workplace is injury to the eyes arising from direct accidental exposure of the eye to chemicals. This kind of injury could result from chemical splashes, etc. Therefore, this section will focus on local effects.

1.3.2.1   *Local effects of chemicals on the eye*   In the workplace, chemical splashes, mist, vapours, gases or dust are probably the most common forms in which chemicals come into contact with the eyes, although there are certainly other methods as well.

Different chemicals can produce different effects. For example, some may cause primary irritation or even corrosive effects to the eye and surrounding structures, whereas others may cause an allergic reaction (allergic conjunctivitis). The severity

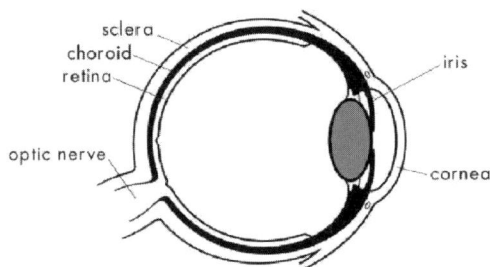

Fig. 1.4   Horizontal cross-section of the human eye.

and time to onset of eye injury will also vary with different chemicals. As examples, consider the effects of alkalis, acids, detergents and solvents. All of these can cause direct corneal damage upon exposure. Strong acids and alkalis will cause damage due to their penetrating effects. For acids, the destructive effects are related both to pH as well as to the affinity of the cation for the corneal tissue. The corneal effects of alkalis, however, are due to the pH value and not to the nature of the cation. Alkalis cause damage to cellular membranes because they saponify the lipids that are present. The result is that further penetration is promoted giving rise to severe lesions. Relative to acids, there tends to be a slower onset of effects and the pH, concentration and duration of exposure to alkalis are the determining factors when considering the severity of the outcome.

Organic solvents act in a similar manner to alkalis although their effects are not usually so penetrating or severe. When considering detergents, the cationic variety is more damaging to the cornea than the non-ionic or anionic variety. Keratitis (inflammation of the cornea) is the most common outcome.

1.3.2.2 *Why test for eye irritancy and corrosivity?*   Accidental eye injuries are very common not just in the workplace, but also in the home. Even taking into consideration the fact that ocular toxicity or injury rarely causes death, it is a very emotive issue because, for the vast majority, the thought of blindness is a terrifying prospect. Therefore, there is a need to know of any potential eye hazards when handling chemicals. Labelling can then be used to warn the user. Advice can be provided on the most appropriate personal protection and the first aid measures to use should exposure occur.

In 1944 Draize *et al.* formalised a test protocol for eye irritation/corrosion testing and since then it has been extensively used. However, different regulatory bodies can use different types of schemes for the interpretation of test data and the ultimate classification of results. The OECD 405 test method for acute eye irritation/corrosion is outlined in the following section.

1.3.2.3 *Acute eye irritation/corrosion studies (OECD 405)*   The OECD 405 test guideline defines eye irritation as 'The production of reversible changes in the eye following the application of a test substance to the anterior surface of the eye'.

This test involves the application (0.1 ml) of the test compound directly into the eye of an albino rabbit, with the other eye used as a control. After 24 hours any remaining traces are washed out, or sooner if necessary. Four animals are usually used and the effects are graded up to 72 hours following exposure. The reactions of the conjunctiva, cornea and iris are visually scored, using the scheme provided in the test method.

Many of us have experienced the discomfort and pain from getting shampoo in the eyes and can almost certainly understand why there are so many objections based on purely ethical reasons to such a test. Therefore, in order to reduce any potential animal suffering, those chemicals that have been shown to be either severely irritating or corrosive in dermal irritancy/corrosion studies, do not have to be tested. Instead, they are automatically labelled in the same way, the view being taken that because the membranes of the eye are, by far, more delicate than the skin, then the reaction would almost certainly be the same, if not worse! Similarly, those chemicals with a pH ≤ 2.5 or ≥ 11.5 need not be tested. Instead, they immediately become classified as corrosive and any results from well-validated *in vitro* studies (see below) which show the chemical to be either potentially severely irritating or corrosive also need not be done. Therefore, if it is necessary to have this kind of information for a chemical, then it is better to test for skin irritancy first and then, based on the results, decide whether or not it is necessary to test for eye irritation. This is also recommended in the OECD guidelines and it avoids the issue of having to do unnecessary animal tests and of course, even reduces the costs, which should please the company accountant!

1.3.2.4 In vitro *tests* Considerable research has been carried out in the area of finding alternatives to animal testing and two of the more successful approaches developed have been with irritation/corrosion studies, where the drive has essentially been to replace animals used to grade irritation with an alternative method. The alternative studies are called *in vitro* testing, which is Latin for 'in glass', and *in vivo* testing, meaning 'in the body', i.e. animal tests. These tests have been validated for use with certain irritants, namely those which are suspected of being severely irritating or corrosive. (Validated means that the test method used produces similar results when different test laboratories carry them out.) Unfortunately, these tests cannot be used for proving the lack of irritancy or for grading mild/moderate irritants. In these cases, the actual animal study needs to be done. But at least it can be said that these tests are a step in the right direction with regard to replacing the use of animals.

1.3.3 *Respiratory irritation*

During the process of breathing, the respiratory tract is exposed to chemicals which are present in the air and which may be inhaled. Therefore, it is not surprising that in the industrial setting, inhalation is the second most common route of exposure.

The upper respiratory system comprises the nose, pharynx, larynx and the trachea. The trachea divides into two bronchi which upon entering the lungs, sub-divide into bronchioles. The bronchioles then divide into alveolar ducts, which terminate as alveoli (Fig. 1.5). An adult human being has many millions of alveoli and their function is gaseous exchange. Atmospheric oxygen is absorbed and carbon dioxide is eliminated.

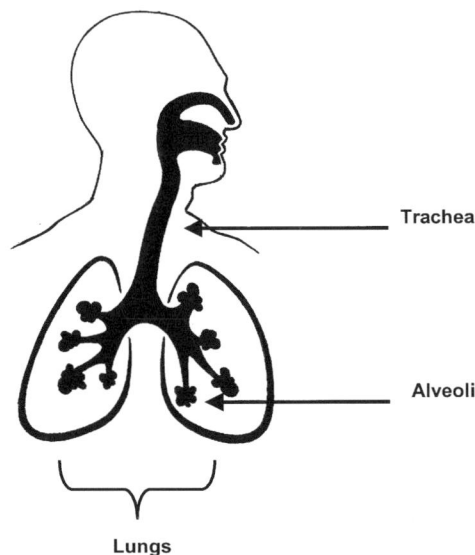

Fig. 1.5   The respiratory tract.

1.3.3.1   *What can be inhaled?*   Although gases, vapours, and solid particles are inhalable, size, shape and water solubility are the main physical factors which will determine where within the respiratory tract they will be deposited. For particulate material, size and shape are the important factors in determining where within the respiratory tract it will be deposited. Overall, deposition somewhere within the respiratory tract is likely to occur for particles in the size range of 0.01–10 µm. Water-soluble gases such as sulphur dioxide tend to be deposited in the upper regions of the respiratory tract.

   Although many airborne contaminants are coughed out, some can stay in the lungs where they can cause either local or systemic effects, i.e. an airborne contaminant can be inhaled and, if upon reaching the lungs, it could be absorbed and transported away via the blood stream to other parts of the body. Otherwise, it could cause local damage to the lungs. This could include eliciting an allergic reaction such as asthma. Alternatively the contaminant could cause pulmonary irritation.

1.3.3.2   *Pulmonary irritation*   When considering inhaled respiratory irritants, there are various factors that will determine the extent and site of injury. These include the size, shape and concentration, along with the extent and duration of exposure. In most cases, an inflammatory reaction is the most common outcome and it is the upper respiratory tract, which is most commonly affected. Chlorine and ammonia gases are classic respiratory irritants.

1.3.3.3   *Testing for respiratory irritation*   Other aspects of respiratory effects, such as toxicity and allergic effects, are covered in other sections of the book. When considering respiratory irritation, there is no specific animal test available for the detection of chemical substances that cause such effects. Instead, the OECD guideline number 403 'Acute Inhalation Toxicity' is usually used. One dose level is usually employed and upon termination of the study the tissues of the respiratory tract are examined for any signs of inflammation.

Reference

Hunter, J.A.A. & Aldridge, R.D. (no date) *Mechanisms of Skin Irritation and Allergy.*

Further reading

Percutaneous Absorption
Monograph No. 20
ECETOC, 1993

Eye Irritation Testing
Monograph No. 11
ECETOC, 1988

Skin Irritation
Monograph No. 15
ECETOC, 1990

Occupational Toxicology
Edited by Neill H. Stacey
Taylor & Francis, 1993

Basic Toxicology: Fundamentals, Target Organs and Risk Assessment, 3rd edn
Frank C. Lu
Taylor & Francis, 1996

General and Applied Toxicology, abridged edn
Edited by Bryan Ballantyne, Timothy Marrs & Paul Turner
Macmillan Press, 1995

Dermal Toxicity
J.H. Draize [In USA *Appraisal of the Safety of Chemicals in Foods, Drugs and Cosmetics*, Edited by Editorial Committee of the Association of Food and Drug Officials of the United States, Austin, TX]
Texas State Department of Health, 1959

OECD Guideline for Testing of Chemicals
Acute Inhalation Toxicity, No. 403

OECD Guideline for Testing of Chemicals
Acute Dermal Irritation/Corrosion, No. 404

OECD Guideline for Testing of Chemicals
Acute Eye Irritation/Corrosion, No. 405

1.4   Chemical allergies

1.4.1   *The immune system*

We are constantly exposed to a hostile environment of microbes that could give rise to diseases. It is the immune system that our bodies utilise in order to ward off such intruders. The immune system provides two kinds of protection, both a non-specific, innate immunity and a specific, acquired or learned immunity. Both of these systems are intertwined in their function. However, although the immune system is our mechanism to ward off micro-organisms, the protective effects of the immune response can often lead to tissue damage (more about this later).

For the purpose of this section, attention will be given to the acquired immune response.

1.4.1.1 *What causes an immune response?* The immune response is activated as a result of contact with an antigen. As will be seen, it can take many forms, such as hypersensitivity, antibody production, cell-mediated immunity, etc.

An *antigen* is simply a substance that is recognised as foreign by the body and which is capable of initiating an immune response. It could be anything from a micro-organism (virus, bacterium), both of which are capable of causing disease, to other substances such as chemicals or pollen. Many people know of the misery caused by pollen during the summer months in the form of hay fever.

There are three important features of the acquired immune system. Firstly, the immune response is *specific* for the type of antigen it encounters. Secondly, it has the ability to remember previous encounters with specific antigens, thereby allowing a more rapid response on subsequent exposures and prevents the disease from having the chance to take a hold (this is the principle of immunisation). That is, it has *memory*. Thirdly, it can *recognise* self from non-self (until things go wrong!).

1.4.1.2 *How do the lymphocytes play a role in the immune response?* The organs of the immune system are situated throughout the body and are often called lymphoid tissues because they are involved with the growth, development and deployment of the lymphocytes (white cells) There are two main populations of lymphocytes involved in the immune response. These are the T-lymphocytes (often known as T-cells) and the B-lymphocytes (B-cells).

The immune response can be mediated by either the B-lymphocytes (B-cells), which is called *antibody-mediated immune response*, and which involves the production of immunoglobulins (or what are more commonly known as antibodies, examples of which include IgE, IgM and IgG), otherwise it can be mediated by the T-lymphocytes (T-cells), in which case it is known as a *cell-mediated immune response*. Of course, the type of response will depend on the type of foreign substance encountered, although it is not unusual for both these types of response to be triggered.

1.4.2 *Chemical hypersensitivity*

An allergy is an adverse reaction caused by an over-stimulation of the immune system in response to a specific allergen that is otherwise harmless and would normally be tolerated by the majority of those who come into contact with it. It is the clinical manifestation of hypersensitivity.

Many of us have probably had a form of allergy at some time or another in our lives, hay fever (pollen allergy) is a very common example. Different types of allergies will give rise to different symptoms depending on the tissues of the body involved.

These symptoms can include sneezing, runny nose and eyes, coughing and skin rashes.

A word commonly seen in print is 'allergen'. This is simply a foreign substance (an antigen) that causes an allergic reaction in a hypersensitive person. Typical allergens are proteins such as pollen, peanuts, shellfish, etc. Allergens are either:

- *Whole allergens* – allergens able to elicit an immune response themselves. Complex biological substances such as pollen typically act as whole allergens.
- *Haptens* – these allergens themselves do not trigger an immune response. Instead, they combine with and modify the body's own proteins to form complete antigens. It is this combination that the body sees as 'foreign'. Many chemicals, and in particular those which are of a low molecular weight, will act as haptens.

1.4.2.1 *How does an allergic reaction occur?* An allergy is an adverse effect resulting from specific immunological responses and it has two stages.

- *Sensitisation* (or induction) *phase* – during the first encounter with the allergen, the individual becomes sensitised as a result of changes in the immune system which involve the lymphocytes.
- *Elicitation phase* – subsequent exposure to the same allergen, which can be at much lower concentrations than experienced in the sensitisation phase, can lead to the development of allergic symptoms.

In a cell-mediated response, there is generally a delay of between 24 and 48 hours following exposure before symptoms develop in a sensitised individual. This is often why it is called a *delayed response*. An example of this is contact hypersensitivity, which could be caused, for example, by formaldehyde. However, with an antibody-mediated response the symptoms are seen within minutes of exposure (again in a sensitised individual), which is why it is often called an *immediate response*. Occupational asthma is an example of this.

1.4.3 *Contact hypersensitivity (allergic contact dermatitis)*

1.4.3.1 *What is allergic contact dermatitis?* Allergic contact dermatitis (another name for skin sensitisation) arises as a result of the skin becoming sensitised to a specific allergen. Subsequent exposure to that same allergen results in an inflammatory reaction.

Of particular interest to us is the handling of chemicals used during the pulp and paper-making process. As may be recalled, most chemicals, and especially those that are of a low molecular weight, are unable to provoke an immune response on their own. This is because they are too small to be recognised by the immune system. Instead, they act as haptens. Common contact allergens include formaldehyde, epoxy resin and nickel.

1.4.3.2 *What is the mechanism of allergic contact dermatitis?* As is the case with allergy, there are two stages in the development of allergic contact dermatitis.

(1) *Induction/sensitisation.* The chemical substance penetrates the skin where it combines with natural proteins in the epidermis thereby forming a complete antigen. This is subsequently transported to the local draining lymph nodes where an immune response is stimulated. The whole process usually takes at least 10 days, after which time the individual is sensitised (allergic). During this induction/sensitisation stage there are no indications of skin damage.

(2) *Elicitation.* Subsequent contact elicits an immune reaction with the outbreak of dermatitis usually within 48 hours. The symptoms are that of an allergic reaction, i.e. itching, redness, pain, rash, etc.

1.4.3.3 *What factors dictate whether or not allergic contact dermatitis will develop?* Although there is no way to predict who will develop allergic contact dermatitis, there are some factors which can play a role. These include the following.

(1) Any predisposing skin condition such as irritant contact dermatitis or simply skin abrasions or cuts will increase the possibility of developing allergic contact dermatitis. This is because any skin damage makes it much easier for chemicals to enter the skin layers thus allowing sensitisation to occur more readily.

(2) As with irritant contact dermatitis, the working environment will be of importance and the same principles will apply (sweating, friction, etc.).

(3) The ability of the chemical to penetrate the upper layers of the skin is an important factor and its allergenic ability will depend upon this. Thus, if it cannot penetrate the upper layers of the skin, then it will not be able to invoke an immune response, and therefore, there will be no allergic reaction. The duration and frequency of exposure together with the concentration and amount of chemical allergen involved will also play a role. A weak allergen would require more frequent contact compared to a strong allergen before any sensitisation effects were induced.

(4) Genetic predisposition (atopy) will also be a factor.

However, not all the workers who handle a chemical that has potentially allergenic properties will develop this disease. In some cases, workers could have worked with such a substance for many years before any symptoms occur. For others, it could develop much more rapidly. Similarly, although it is possible to become sensitised to a particular allergen, it is not always the case that an outbreak of dermatitis will occur upon subsequent exposures.

1.4.3.4 *What are the consequences of developing allergic contact dermatitis?* The problem with allergic contact dermatitis is that it can persist for many years and, once sensitised, the only way to avoid the outbreak of the symptoms (dermatitis) is to avoid all contact with that specific allergen. However, even if all efforts are made to avoid contact with that specific chemical, it is possible to get cross-sensitivity reactions.

As with other forms of hypersensitivity, allergic contact dermatitis is a very important issue in the workplace because normal handling and protective precautions are often ineffective. Even if a particular chemical has an occupational exposure limit and attempts are made to keep the levels well below this, the breakout of symptoms associated with that particular allergy in sensitised individuals may still occur. This is because the levels needed to trigger an outbreak of symptoms are very low and often well below that specified by an occupational hygiene limit.

1.4.4 *Respiratory hypersensitivity (occupational asthma)*

Asthma is a chronic inflammatory lung disease whereby the airways sporadically constrict in response to some kind of stimuli. This results in bronchoconstriction and inflammatory effects that make breathing difficult, eventually leading to asthma. Asthma attacks are frequently serious and can sometimes even be fatal. It has been estimated that between 2 and 15% of all asthma cases world-wide are occupational in origin, and it appears that the figures are rising (Blank, 1987), although the reason for this rise is not clear.

1.4.4.1 *What is occupational asthma?* Occupational asthma is the most important clinical manifestation of respiratory hypersensitivity in the chemical industry (Magnusson & Kligman, 1969). It is where the airways of the lung in a susceptible individual overreact to a substance found in the workplace that would normally be tolerated by the majority of the working population. It is the clinical manifestation of respiratory sensitisation. Exposure can either induce asthma symptoms in these individuals who have previously never suffered from the disease, or it can trigger a pre-existing asthmatic condition. Typically, asthmatic symptoms due to occupational exposure are seen to improve when the individual is away from the workplace, such as at weekends and during vacations, only to reappear upon return to work.

1.4.4.2 *Which occupations can give rise to occupational asthma?* Although the incidence of occupational asthma can vary within different professions, there are some working environments which are well-known 'trouble-spots' with respect to this condition. In particular, those working in the bakery or detergent industries appear to be at risk. It has been suggested that up to 9% of all bakers suffer from occupational asthma (Bakers' asthma) as a result of hypersensitivity to cereal proteins or insect contaminants (Theil & Ulmer, 1990). Within the detergent industry, it has been estimated that up to 45% of the work-force may develop respiratory symptoms as a result of exposure to one of the enzymes used in washing powder (Blank, 1987).

A further problem is topical allergy. Latex allergy is a condition that has emerged over the past decade in health care workers with the need to protect against blood-borne diseases by the wearing of latex gloves. The problems arise when susceptible workers become sensitised to the aerosolised latex protein allergens. This has become a major occupational health care problem.

1.4.4.3 *Mechanisms of occupational asthma* Many of the substances causing occupational asthma are often very common and are usually considered harmless. (This can be seen from the three examples mentioned above.) There are, in fact, over 200 substances which have been implicated in the development of occupational asthma (Chan-Yeung & Lam, 1986). These substances can either cause occupational asthma as a result of interaction with the immune system (immunological), or otherwise via a non-immunological mechanism, as can be the case with certain irritants (Fig. 1.6).

```
┌─────────────────────┐      Occupational       ┌─────────────────────┐
│   Immunological     │ ───▶                  ◀── │        Non          │
│    Mechanism        │        Asthma            │   Immunological     │
│                     │                          │     Mechanism       │
└─────────────────────┘                          └─────────────────────┘
```

Fig. 1.6 Mechanisms of occupational asthma.

1.4.4.4 *Non-immunological mechanism* This is also known as 'irritant-induced asthma' or 'reactive airways dysfunction syndrome'. In this case, exposure to very high levels of irritants, such as ammonia or chlorine gas, as is the case during some form of industrial accident, can on occasion, cause symptoms of asthma. Workers who have a pre-existing asthmatic condition or another respiratory disorder are particularly susceptible to this kind of exposure. Since this form of occupational asthma does not involve the immune system, there is no latency period, which of course means that the symptoms usually develop rapidly (within 24 hours) and there is no extreme sensitivity to low levels of these irritants (as they are non-sensitising). Therefore, exposure to so-called 'normal levels' of these irritants would not give rise to asthmatic symptoms.

1.4.4.5 *Immunological mechanism* This form of asthma is probably of more interest to those who handle chemicals in the workplace. Bakers' asthma and latex allergies are two examples of immunologically mediated occupational asthma. As with other immunologically mediated diseases, there is a latency period, which in this case can be anything up to three years, after which time the individual is sensitised to the offending substance. Rhinitis and Conjunctivitis (runny nose and eyes) are the most common initial symptoms of sensitisation which can develop into asthma should exposure continue.

It is well known that substances (or more correctly, allergens), which cause this kind of asthmatic response in susceptible individuals, are either high molecular weight compounds or they are low molecular weight reactive compounds. A few examples are given in Table 1.3. Although many workers may be exposed to the same allergen, only a relatively small proportion may go on to develop clinical symptoms of asthma. This of course leads to the question, 'who is most susceptible?'

In order to attempt to answer this question, a number of studies have been carried out. The results indicate that there are certain host factors that are involved. One of them is atopy. An individual is said to be atopic if he or she has the natural tendency to produce specific IgE antibodies in response to certain allergens that are present in the surroundings. It has been estimated that atopy is in fact present in around 30% of the general population.

Table 1.3 Common compounds that can cause an asthmatic response in susceptible individuals.

High molecular weight compounds	Typically complex biological substances such as wheat flour, rodent proteins, latex, antibiotics, etc.
Low molecular weight compounds	Industrial chemicals such as anhydrides, epoxy resins, diisocyanates, platinum salts, etc.

Research has shown that high molecular weight compounds initiate an antibody response which usually involves the production of specific IgE antibodies. Furthermore, studies have indicated that atopy can be a predisposing factor when individuals are exposed to high molecular weight compounds (Chan-Yeung, 1990). This of course makes sense in light of what is known about immunologically mediated occupational asthma which has been triggered by high molecular weight compounds.

However, it is a different picture with low molecular weight compounds. The immune response in this instance involves a more complicated mechanism, not necessarily involving specific antibodies. With these compounds, atopy does not appear to be such an issue.

1.4.4.6 *Multiple exposures – the 'real-life' working situation* Many of us who have worked in a chemical laboratory already know that it is very common for different types of chemicals to be handled at any one time. This of course means that exposure is unlikely to be from one chemical alone and that in some cases concurrent exposure could enhance the hazardous effects seen.

For example, it is known that concurrent exposure to respiratory irritants in the workplace can increase the probability of susceptible individuals becoming sensitised to other chemicals (Osebold *et al.*, 1980). This is particularly so within the platinum industry where it has been observed that cigarette smokers are more susceptible to the development of asthmatic symptoms than non-smokers. Cigarette smoke is not in itself a sensitiser, but instead acts as a respiratory irritant. What is believed to happen is that the smoke irritates the bronchial mucosal layer, thereby allowing platinum allergens easy access. In other words, the smoke acts as an adjuvant (Venables *et al.*, 1989).

Although some of the industries mentioned are not related in any way whatsoever to the pulp and paper industry, they are still of importance because they highlight the potential adverse health effects that can occur when handling different chemicals. This provides a golden opportunity to improve our own safety measures currently used in occupational settings. It has been said that a wise person is one who learns from someone else's mistakes!

Occupational asthma does not simply 'go away'. Once individuals have become sensitised to a substance in the workplace, it stays with them. Such that even if exposure is stopped and the symptoms disappear completely, they can reappear, even several years later, should exposure recur. In fact, within the workplace, any continued exposure to irritant levels that can be much lower than any threshold limit value (if it exists) can aggravate the condition. Therefore, for those who are unfortunate enough to suffer from asthma brought on from their working environment find that the impact on their lives can be severe.

Seeking an alternative form of employment is often the only solution to avoiding further asthmatic attacks, which can have adverse economic consequences for both the individual and even the employer.

1.4.5   *How to test for allergic reactions*

There are two exposure routes that need to be considered: skin contact and inhalation.

1.4.5.1   *Skin sensitisation studies (OECD 406)*   Two OECD test methods are most commonly used when investigating skin sensitisation. These are the Buehler test and the Guinea Pig Maximisation Test (GPMT). The only difference between these two tests is that the GPMT involves the subcutaneous injection of the test compound and the use of an adjuvant, whereas the Buehler test involves the topical application of the test compound without the adjuvant. (An adjuvant is a substance which is used to enhance or modify the immune response in a non-specific manner.)

Both procedures essentially involve exposing the test animal (guinea pig) to the chemical being analysed by either subcutaneous injection or topical application. After an interim period of at least seven days (whereupon an immune response may or may not have developed) the animals are exposed to a challenge dose and any allergic symptoms are noted. These are compared to a control group and graded according to their severity. Some MSDS sometimes use terminology such as 'strong sensitiser', 'weak sensitiser', etc. in an attempt to grade the allergenic potential of chemicals that have been tested as per the GPMT method (see Table 1.4). This grading is based on the percentage of animals sensitised during the test and has no meaning whatsoever for labelling and classification purposes.

Table 1.4   The grading as used in the Maximisation Method.

Sensitisation rate (%)	Classification
0–8	Weak
9–28	Mild
29–64	Moderate
65–80	Strong
81–100	Extreme

Source: Magnusson and Kligman (1969).

For classification and labelling purposes in the European Union and Canada, at least 30% of the test animals have to demonstrate a response in an adjuvant-type test (the GPMT) in order for the test compound to be classified as being a skin sensitiser. For a non-adjuvant-type test, however, the figure is lower with a 15% response at least required. Though many in the scientific community consider the GPMT to be unrealistic because it involves the injection of the test material and the use of an adjuvant it does remain the test which is most accepted and requested by regulatory bodies in the classification of skin sensitisers.

1.4.5.2   *Testing for respiratory sensitisation*   At present, no widely accepted tests for respiratory sensitisation have been recognised. There are two tests that can be used, but these are specific to pulmonary responses in guinea pigs and IgE measurements in mice.

Therefore, the classification for respiratory sensitisation comes mostly from human experience and in some cases, the use of structure–activity relationships. It has been reported that generally chemicals not having the potential to cause allergic contact dermatitis are also unable to induce respiratory allergy (Briatico-Vangosa *et al.*, 1994).

References

Blank, P. (1987) Occupational asthma in a national disability survey. *Chest*, **92**, 613–617.

Briatico-Vangosa, G., *et al* (1994) Respiratory allergy: hazard identification and risk assessment. *Fundamental and Applied Toxicology*, **23**, 145–158.

Chan-Yeung, M. (1990) Occupational asthma workshop on environmental and occupational asthma. *Chest*, **98** (Suppl.), 145S.

Chan-Yeung, M. & Lam, S. (1986) Occupational asthma – state-of-art. *American Review of Respiratory Diseases*, **133**, 688.

Magnusson, B. & Kligman, A.M. (1969) The identification of contact allergens by animal assay. The Guinea Pig Maximization Test. *Journal of Investigations in Dermatology*, **52**, 268.

Osebold, J.W., Gershwin, L.J. & Zee, Y.C. (1980) Studies on the enhancement of allergic lung sensitization by inhalation of ozone and sulphuric acid aerosol. *Journal of Environmental and Pathological Toxicology*, **3**, 221–234.

Theil, H. & Ulmer, W.T. (1990) Bakers' asthma: development and possibility for treatment. *Chest*, **78**, 400.

Venables, K.M., Dally, M.B., Nunn, A.J., *et al.* (1989) Smoking and occupational allergy in workers in a platinum refinery. *British Medical Journal*, **299**, 939–942.

Further reading

Occupational Toxicology
Edited by Neill H. Stacey
Taylor & Francis, 1993

General and Applied Toxicology, abridged edn
Edited by Bryan Ballantyne, Timothy Marrs & Paul Turner
Macmillan Press, 1995

Basic Toxicology: Fundamentals, Target Organs and Risk Assessment, 3rd edn
Frank C. Lu
Taylor & Francis, 1996

Skin Sensitisation Testing
Monograph No. 14
ECETOC, 1990

Respiratory Allergy
Monograph No. 19
ECETOC, 1993

OECD Guideline for Testing of Chemicals
Skin Sensitisation, No. 406

Dermatoxicology Test Techniques: An Overview
Marshall Steinberg
Raven Press, 1984

Cutaneous Toxicity
Edited by V.A. Drill & P. Lazar
Raven Press, 1984

Variable effects of chemical allergens on serum IgE concentration in mice. Preliminary evaluation of a novel approach to the identification of respiratory sensitisers. *Journal of Applied Toxicology*, **12** (1992) 312–332
R.J. Dearman, D.A. Basketter & I. Kimber

Principles and Practice of Immunotoxicology
Edited by K. Miller, J. Turk & S. Nicklin
Blackwell Science, 1992

1.5   Genetic toxicology and carcinogenicity

The role of genetic toxicology is to identify chemicals that have the potential to cause alterations in the genetic material of our cells which could then be passed on to our offspring. Such chemicals are called genotoxins.

It is a relatively 'new kid on the block' compared to other areas of toxicology. Testing for potential genotoxic effects of new chemicals was only made mandatory by the majority of regulatory agencies as recently as the 1970s. This followed growing concern over the exposure of workers to industrial chemicals that could possibly cause heritable genetic alterations (mutations).

Genetic toxicity testing aims primarily to identify those chemicals which have the potential to cause heritable effects (mutations) in humans and also to pin-point potential carcinogens.

Although mutagenicity and carcinogenicity are different disease processes, they both have the similar feature of being able to induce genetic alterations in cells. In order to understand how genetic alterations can arise, a little background information is outlined below concerning the genetic material of our cells and cell replication.

1.5.1   *The cell and its genetic material*

The cell is the basic functional unit of all living organisms and carries genetic information in the form of *chromosomes*. These are a type of organelle present in the cells of eukaryotes (plants and animals). Humans have 23 chromosome pairs, or 46 chromosomes, in non-reproductive cells, while reproductive cells (sperm and ova) contain 23 chromosomes, i.e. half the number. The main constituent of the chromosome is deoxyribonucleic acid (DNA), which is a very large polymer that makes up the primary genetic material of cells. DNA comprises smaller repeating units called nucleotides (or bases) of which there are four main types: adenine, guanine, cytosine and thymine. Together they can form many different sequences.

The genetic information, or *genetic code*, which is carried by the chromosomes is contained in its DNA, where it is encoded in the different sequences of DNA nucleotides. The genetic code can be viewed as being a set of instructions for the construction of specific proteins and enzymes.

So, where do *genes* fit into the picture? Many who have heard of the term 'genes' may have done so in the context of a physical trait, such as hair colour or eye colour inherited from a parent. Genes are the smallest part of the DNA and they code for a heritable trait or function in an organism, i.e. they determine a particular characteristic of an individual. Humans have many different types of genes and in turn many different types of chromosome which gives rise to different traits that makes us what we are.

1.5.2 *Cell replication – mitosis and meiosis*

From time to time new non-reproductive (somatic) cells will be needed in order for an organism to either replace damaged or dead cells, or simply to grow. Such cells are produced by *mitosis* during which the DNA replicates along with the chromosomes and the cell divides into two new daughter cells which are genetically identical.

Meiosis creates new gametes and the process occurs in the gonads of animals. Gametes are mature sex cells, i.e. sperm cells in the male and ova or egg cells in the female. During meiosis the chromosomes of the original cell split into two which means that each of the resulting new cells contains half the original number of chromosomes – they are known as haploid cells. During the sexual reproductive process, an egg and sperm cell can combine so as to produce a new cell which contains the full number of chromosomes (a diploid cell) and this cell can go on to develop into a new individual.

During the process of cell division, it is absolutely essential that replication of the genetic information is completely accurate in order to ensure that it is transferred unchanged to the new daughter cells. Any changes arising from mistakes in the cell division process can have dire consequences as will be seen later.

1.5.3 *Mutagenicity*

1.5.3.1 *What is a mutation?*

The OECD guidelines defines mutation as 'A change in the information content of the genetic material which is propagated through subsequent generations of cells or individuals'.

Mutations can either arise spontaneously or can be induced by either chemicals or physical agents such as X-rays. The result is that the offspring differs unexpectedly from the parent that has been exposed to the mutagen. (A mutagen is an agent which is capable of inducing a mutation.)

1.5.3.2 *How do mutations occur?*

Anything that has the ability to cause an alteration in the genetic material of the cell which can then be replicated is called a mutagen and as such it may produce a mutation. The alteration can be within:

- individual genes, whereby a wrong base pair is substituted, deleted or inserted into DNA; once such an error is incurred, it can be perpetuated by further DNA replications;
- chromosomes, whereby the structure or number is altered – any change in structure will normally result in loss of genetic material or in new gene arrangements; a gain in the number of individual chromosomes can result in sterility because fertile gametes are unable to be formed.

The good news is that not all exposures to chemical mutagens will give rise to a mutagenic event. This is because the cells of our bodies contain special biological repair systems which, putting it in a very simplified manner, can read the genetic code and repair any inconsistencies which are found. If this repair is error-free, then the code is restored to the original. Unfortunately, the repair process is not always 100% accurate and in some cases mistakes can occur (known as an 'error prone' process), which could give rise to a mutation should the cell have the ability to undergo successful cell division.

There is one genetic disorder which occurs in humans and highlights the important role played by these DNA repair systems. It is known as Xeroderma Pigmentosa and arises as a consequence of a deficiency in the DNA repair system which renders an individual more susceptible to skin cancer caused by ultraviolet (UV) light.

1.5.3.3 *What is the significance of mutations?* As can probably be recalled, a mutagen causes a permanent change in the DNA coding (genetic code) thereby compromising both the ability of DNA to replicate and carry information. This can have different implications depending on whether the mutation occurs in the *germ* cell (reproductive cells such as sperm or egg) or the *somatic* cell (non-reproductive cells).

Germ cell mutations are of significance because they are inheritable. That is, if the affected cell replicates, the altered DNA coding can be passed on to the progeny. This in itself means that for the particular species involved, the gene pool could ultimately be altered, although of course it would take a number of generations before any effects manifested themselves. Between 5 and 10% of germ-cell mutations can occur in children and Down's syndrome, which is a result of abnormal chromosomal numbers, is an example. Similarly, some spontaneous abortions and still-births have also been attributed to changes in genetic information, i.e. mutations. In contrast to germ-cell mutations, *somatic cell* mutations cannot be inherited by the offspring. When such a mutation arises, there are various scenarios that can occur. For instance, the change arising from the mutation would be of no importance should the cell either die or simply fail to divide. However, should the mutation occur in a cell undergoing division then daughter cells could be formed with this alteration and could die, or live and be capable of dividing into new cells. As a consequence of this last scenario, there is a risk that a somatic mutation (if there were sufficient numbers of cells in the same area with the same mutation) could lead to the formation of a new cell-line which could give rise to a tumour.

It is accepted widely within scientific circles that for a number of chemical carcinogens this kind of mutation is the initiating event in the carcinogenic process.

1.5.4 *Carcinogenicity*

Although both man-made and naturally occurring chemicals can have many adverse (and even fatal) health effects, the effect of most concern to many is cancer. When considering that in many parts of the world cancer is the second most frequent cause of death, this concern is not surprising. Cancer is not a new phenomenon, and nor is the implication of chemicals in being a factor involved in the aetiology of this dreadful disease. In fact, as far back as the eighteenth century, certain occupations (although it was in fact chemicals) were recognised as aetiological factors in human carcinogenesis. As a result of the public perception of cancer, chemical carcinogens are subject to special laws and regulations which essentially restrict their use and thereby human exposure to them (see Chapter 4).

Unfortunately, in the field of cancer, there is a lot of jargon, much of which can be regularly seen in Safety Data Sheets/Material Safety Data Sheets and other health and safety documentation. Throughout this section, reference will be made to the most common terms and, of course, to what they actually mean.

1.5.4.1 *What is cancer?* The *Collins English Dictionary* (1991) defines *cancer* as 'Any type of malignant growth or tumour, caused by abnormal and uncontrolled cell division'.

The term *carcinogenesis* is derived from the Greek word 'karkinos' – which means crab-like – and 'gennan', which means to produce. Thus, *carcinogenesis* means the creation of cancer and *carcinogens* are those agents, which cause cancer. The word *neoplasm* is commonly seen in literature too. It is derived from 'neo' (new), and 'plasm' (growth). It is simply another word for tumour or growth. Throughout this section these words will be interchanged with each other.

Cellular proliferations that result in neoplasms are classified as *benign* or *malignant* and also as to the cell type from which they originate, although, in some cases, it can be very difficult to judge whether the tumour is benign or malignant in nature.

A *benign neoplasm* is typically localised and non-invasive. It does not metastase, i.e. it does not spread to other parts of the body. A *malignant neoplasm* on the other hand is invasive, non-structured and it has the propensity to metastase.

1.5.4.2 *How cancer occurs* Chemical carcinogenesis is a complicated multi-staged process that requires multiple, independent genetic changes to take place before the transformation of a normal cell into one which is cancerous can occur. This process can be theoretically divided into three main stages: initiation, promotion and progression/conversion (see Fig. 1.7).

Since so many processes need to occur before a cell becomes cancerous there is often a long *latency* period before a tumour is observed. What this means is that the clinical symptoms of cancer often appear later on in life, long after the causative agent has disappeared. This can be highlighted with cigarette smoking where it can take up to 20 years before a malignancy develops. (This is probably the reason cancer is often called 'the disease of old age'.)

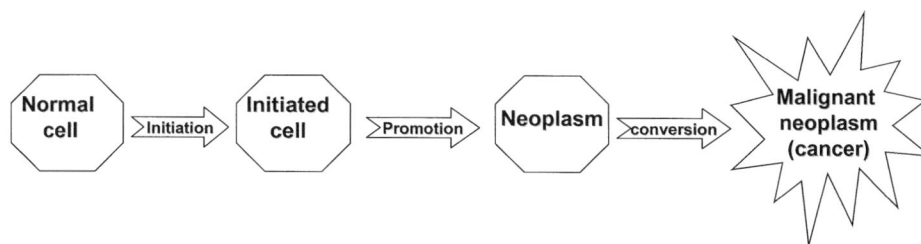

Fig. 1.7 The multiple-staged process of chemical carcingenesis.

1.5.4.3 *Categories of chemical carcinogens*   There are two different categories of carcinogen based on their mode of action and the chemicals that can cause cancer fall into one or other of these categories. These categories are genotoxic carcinogens and non-genotoxic carcinogens (or epigenetic carcinogens).

1.5.4.4 *Genotoxic chemicals*   As a result of their high reactivity, genotoxic chemicals will readily bind to macromolecules, including DNA, and cause some kind of damage. If DNA binding occurs then a mutation may be the result, but this may not persist should either error-free DNA repair remove it or the cell dies. However, should the mutation persist and the cell is able to undergo cell division, then this mutation will remain. The cell containing this mutation becomes what is known as 'initiated', which is believed to be the first step in the process of chemical carcinogenesis.

There are two types of genotoxic carcinogen and two more terms which are often seen in literature.

(1)   *Ultimate carcinogens* – these are direct acting because they possess the correct chemical structure to immediately interact with DNA.
(2)   *Procarcinogens/precarcinogens* – these require some kind of structural conversion before they are able to interact with DNA. The whole conversion process is known as *bioactivation*. (Most chemicals are procarcinogens.)

1.5.4.5 *Non-genotoxic carcinogens (or epigenetic carcinogens)*   Non-genotoxic carcinogens do not cause damage to the DNA. Instead they can either enhance a pre-existing carcinogenic process, or else induce cancer as a result of other cellular effects. Well-known examples of chemicals that act in this manner include hormones, and peroxisome proliferators such as phthalate esters.

1.5.4.6 *Threshold levels*   This is an area of great debate – whether or not there is a threshold level for carcinogens. It is generally accepted that for genotoxic carcinogens there is no threshold level below which exposure will not produce a cancer. In other words an adverse effect level (NOEL) does not exist. However, unlike genotoxic carcinogens, it is believed that a threshold level for non-genotoxic carcinogens does exist. This is because cancer is induced indirectly as a consequence of other cellular effects, i.e. cellular proliferation which acts via a specific mechanism and will exhibit a threshold dose level. Regulatory agencies do not make a distinction between these two types (genotoxic and non-genotoxic carcinogens).

1.5.4.7 *What is known about the causes of cancer?*   There is a variety of causative agents implicated in the development of cancer and it is widely believed that lifestyle, i.e. the use of tobacco products, alcohol, diet, sunbathing, etc., is one of the most important factors involved.

One of the major problems with trying to define the cause of cancer is the relatively long latency period of the disease. This latency period (or induction period) is simply the time between the initiation of the carcinogenic process by a carcinogen to when the clinical symptoms of a pre-malignant or malignant tumour appear. This means that it is very difficult to backtrack and pin-point the causative agent(s), which may themselves act independently or in combination (which complicates matters even more). Another very important factor implicated in the development of cancer is the exposure to carcinogens in the workplace. As it is estimated that up to 8% of all human cancers are occupational in origin, this factor should be taken very seriously.

The risk of developing cancer can vary tremendously within different populations. This can be partly attributed to differences in genetic make-up, which can make certain individuals more 'susceptible' to the development of a particular cancer. Also, some cancers are more prevalent in certain geographical areas of the world than in others. For example, Japan has a higher incidence of colon cancer compared to the USA. There are also forms of cancer that can be inherited, although it is difficult to rule out entirely external factors which may have contributed to its development in such cases. A well-known example of is breast cancer. It is generally the case that with these forms of inherited cancer the symptoms are seen at a much younger age.

As a result of advances in both technology and medicine, people are on the whole living longer than they used to, which of course means that the incidence of cancer is likely to rise in the future.

1.5.4.8 *Will all exposures to a carcinogen cause cancer?* No, not necessarily. As with all other toxic effects, the amount of exposure, or dose, is critical. Therefore, the higher the exposure, both in terms of concentration of chemical and duration of exposure, the greater the risk of harm to the individual and of course, the greater the number of individuals that will be harmed (especially if the repair mechanisms that all the cells of human bodies possess are overwhelmed so that they cannot function properly).

1.5.5 *Genetic toxicity testing*

Because the consequences of mutation can take several generations to manifest themselves, the importance of tests being available that can quickly detect and classify mutagenic chemicals has become all the more apparent.

Probably many readers typically regard mutagenicity tests in the context of being useful, rapid, short-term screening tests for chemical genotoxic carcinogens. This is not an incorrect assumption. As a result of the association between mutagenicity and carcinogenicity, short-term genetic toxicity tests have been developed with the aim of investigating whether somatic cell mutations can be produced leading to the development of a clinical cancer. Nowadays, therefore, mutagenicity tests are used primarily to detect any chemicals that could potentially cause heritable effects and also to screen for potential carcinogens.

As mentioned above, there are two types of mutation – those involving chromosomes and those involving genes. However, it should be noted that not all chemical agents that cause gene mutations cause chromosomal damage and vice versa. Ideally, a test is needed that covers both types of mutation. Unfortunately, no single test is yet available that can do this. Therefore, batteries of tests are used that are able to detect all possible genetic end-points. The testing approach typically follows a stepwise routine until sufficient information is available to identify any potential hazard in terms of both mutagenic activity and carcinogenic potential. As a general rule, most regulators initially want tests that detect both gene and chromosomal effects *in vitro*. This usually means that a reverse bacterial mutation test and chromosomal aberration test are the first to be performed, although it is not unusual for these to be supplemented by further *in vitro* tests should the preliminary test findings so dictate. If the results of more than one of these initial *in vitro* tests indicate a possible mutagenic event, then *in vivo* tests are carried out in order to confirm the results obtained from *in vitro* studies. Specific tests or detailed test methods are not normally stated in the regulatory requirements.

By virtue of the fact that DNA is such a universal structure, a diverse range of species including micro-organisms, mammalian cells, insects, plants and laboratory animals can be used in the tests alluded to above. As was briefly mentioned earlier in the section on carcinogenicity, most chemicals require some form of metabolic activation before they can react with DNA. When performing *in vitro* tests for genetic toxicity, therefore, it is important to take this behaviour into account. A special mixture comprising metabolically competent cells is usually employed in order to provide metabolic activity in the *in vitro* tests. The *in vitro* tests are usually carried out in both the presence and absence of a metabolic activator.

There are many test methods available. Owing to a limitation of space, however, only those as per the OECD guidelines have been listed with a little more information provided about a few common tests often referred to in health and safety literature. As will be seen, these test methods are designed to detect genetic or chromosomal mutations using either somatic or germ cells. The tests can be done *in vivo* or *in vitro* and either with or without metabolic activation.

1.5.5.1 *Gene mutation tests*

1.5.5.1.1 *Bacterial mutation assays*   Probably the most commonly known test in this category is the 'Ames Test'. This is a reverse mutation assay in that it uses a mutated strain of the bacterium *Salmonella typhimurium*, which is unable to synthesise the amino acid histidine. The test is designed to determine whether a mutation can occur as a result of chemical contact with the gene that controls histidine synthesis. If it does then the bacterium is reverted to its original state and can once more synthesise histidine and the colony grows. Different test strains enable many of the possible mutations that can occur to be detected and the advantages of such assays are that they are relatively easy, inexpensive and rapid to execute. However, since this test is carried out using bacteria, it cannot be conclusively stated that the same effects would be seen in eukaryotic cells. Therefore, in the event of positive findings, a second *in vitro* test is usually performed that uses eukaryotic cells such as fungi, yeast or mammalian cell cultures. (See Table 1.5 for various OECD tests.)

1.5.5.1.2 *Mammalian cell mutation assays (OECD 476)*   The two most commonly used cell-lines in these tests are Chinese hamster ovaries (CHO) and mouse lymphoma cells. They are cultured with the test substance and the colony-forming ability is assessed. These assay types can detect different gene mutations and of

Table 1.5   The OECD genetic toxicity tests.

Assays for gene mutations	*Salmonella typhimurium* reverse mutation (OECD 471)*
	Escherichia coli reverse mutation assay (OECD 472)*
	Gene mutation in mammalian cells in culture (OECD 476)*
	Drosophila sex-linked recessive lethal assay (OECD 477)*
	Gene mutation in *Saccharomyces cerevisiae* (OECD 480)*
	Mouse spot test (OECD 484)
Assays for chromosomal aberrations	*In vitro* cytogenetic assay (OECD 473)*
	In vivo cytogenetic assay (OECD 475)*
	Micronucleus test (OECD 474)*
	Dominant lethal assay (OECD 478)
	Heritable translocation assay (OECD 485)
	Mammalian germ cell cytogenetic assay (OECD 483)
Assays for DNA effects	DNA damage and repair; unscheduled DNA synthesis *in vitro* (OECD 482)*
	Mitotic recombination in *Saccharomyces cerevisiae* (OECD 481)*
	In vitro sister chromatid exchange assay (OECD 479)*

* Indicates that test can be used for carcinogen screening.

course, the use of mammalian cells in such an assay is more relevant to humans than the use of bacteria (for obvious reasons).

1.5.5.2 *Chromosomal mutation tests*   These tests are of importance when considering the fact that chromosomal abnormalities are associated with congenital malformations, spontaneous abortions, etc. Chromosomal breakage can be visually seen when using a microscope and appropriate staining techniques.

1.5.5.2.1 *Rodent chromosomal assay*   The *in vivo* mammalian bone marrow cytogenetic test (OECD 475) uses rodents that are exposed to the test substance via a route which is applicable to the intended use of the product. The animals are then sacrificed at various time intervals (which is established based on the rate of bone marrow production) and the structural aberrations are scored by means of staining and examination. This test detects clastogens and of course allows for inherent factors such as pharmacokinetics, metabolism and repair.

1.5.5.2.2 *Micronucleus test (OECD 474)*   Of all the chromosome *in vivo* mutation tests, this is probably the one which is most commonly seen and uses mice, although use of rats is not uncommon. The formation of micronuclei is taken as an indication of chromosomal breakage and occurs as a result of a nuclear membrane forming around any broken bits of chromosome.

1.5.5.2.3 *DNA damage*   Tests that examine the formation of DNA adducts and DNA breakage tend to be used as research tools and are not done as part of any testing strategy. DNA repair synthesis in non-dividing cells occurs only after DNA damage and indicates the existence of this damage. The most commonly used test for this is *unscheduled DNA synthesis* in rat hepatocytes and this provides an indirect measurement of mutagenicity. However, it should be noted that DNA repair only informs us that damage must have occurred, but that the repair could have been error-free, which would therefore be of no consequence. This test is usually done *in vivo*.

1.5.6   *Testing for chemical carcinogenicity*

There are three methods used for the identification of chemical carcinogens: short-term tests, long-term tests and epidemiology.

1.5.6.1   *Short-term tests*   As was mentioned earlier, the discovery that mutagenicity is linked to carcinogenicity prompted the development of further tests that could be used as screening tools. These tests detect chemicals that produce mutations in the somatic cells by a genotoxic mechanism leading potentially to the development of a clinical cancer. They are useful tests in that they provide screening for candidates that ordinarily would have been submitted for long-term studies. However, all they do is confirm whether or not a mutagenic event occurs during the test conditions specified, and any positive results should be taken as an indication of a possible cancer hazard only. These short-term tests cannot be used to screen chemicals that act by a non-genotoxic mechanism.

1.5.6.2   *Long-term tests*   This kind of assay is the only method of experimentally determining whether there is any carcinogenic potential, especially for non-genotoxic carcinogens that cannot be detected in the short-term test methods.

Three dose levels are usually recommended and 50 rodents (usually to begin with) of each sex per dose group are used. The highest dose level is one, which should produce minimal toxicity. The exposure period depends on the test species used – for rats it is 24 months whereas for mice it is 18 months. Other species such as dogs and primates can be used in some circumstances where a non-rodent species is required. As can almost certainly be appreciated, these tests are very expensive to perform, take a long time to do and use many test animals.

The aim of long-term studies is to see whether or not, under the conditions and dose levels used, there is an increase in the incidence of tumours. Unfortunately, it is not possible to infer conclusively from such studies that a given chemical carcinogen, which, for example, causes liver tumours in the test species used, will necessarily cause the same kind of effect in humans. All that can be interpreted from the results is that there is a qualitative indication of carcinogenicity for that chemical. Similarly, not all chemical carcinogens are carcinogenic in all species and it is for this reason that two species are usually tested: a rodent species initially, and should positive results abound from this, a second non-rodent study would be carried out. It is assumed that positive results in two species is enough of an indicator to conclude that carcinogenic activity would also occur in humans.

1.5.6.3   *Epidemiology*   This is the study of disease in populations, and in the field of cancer it is the only way of categorically providing evidence of carcinogenicity in humans. Such studies are inherently time consuming and, as is often the case, inadequate records, confounding factors, etc., make them very difficult to carry out and interpret.

Epidemiological studies will confirm that there has been an exposure to a carcinogen in a given population, but the results provide no information about the levels at which this exposure occurred. Tests are performed retrospectively, i.e. in response to there being thousands of deaths by cancer, with epidemiology being used to find a likely causal link. Therefore, it is not possible to detect any potential carcinogens until many years after the first exposure to it has occurred. As a consequence, the methods provide no information on the potential hazardous health effects of new chemicals being developed or used. Thus, it is necessary to call upon another kind of assessment strategy, one that can be used so as to determine the intrinsic hazard of chemicals before they are exposed to the population. As mentioned above, there are two types of test – short-term tests and long-term tests.

Reference

Collins English Dictionary (1991), 3rd edn. Harper Collins.

Further reading

General and Applied Toxicology, abridged edn
Edited by Bryan Ballantyne, Timothy Marrs & Paul Turner
Macmillan Press, 1995

Basic Toxicology: Fundamentals, Target Organs and Risk Assessment, 3rd edn
Frank C. Lu
Taylor & Francis, 1996

Carcinogenesis: basic principles. *Drug and Chemical Toxicology*, **19** (1996), 133–148
D.B. Couch

Future approaches to genetic toxicology risk assessment. *Mutation Research*, **365**(1–3) (1996), 191–204
Rosalie K. Elespuru

Human carcinogens so far identified. *Japanese Journal of Cancer Research*, **80** (1989), 795–807
L. Tomatis *et al.*

Cellular and molecular mechanisms of multistep carcinogenesis: relevance to carcinogen risk assessment. *Environmental Health Perspectives*, **76** (1997), 65–70
J.C. Barrett *et al.*

The causes of cancer. *Journal of the National Cancer Institute*, **66** (1981), 197–308
R. Doll & R. Peto

Carcinogenesis. *Hamilton and Hardy's Industrial Toxicology*, 5th edn, chapter 82, pp. 597–608
Stephen Mastorides & Carlos Muro-Cacho

Genetic and cellular basis of multistep carcinogenesis. *Pharmacological Therapy*, **46** (1990), 469–486
J.A. Boyd & J.C. Barrett

Occupational Toxicology
Edited by Neill H. Stacey
Taylor & Francis, 1993

A Contribution to the Strategy for the Identification and Control of Occupational Carcinogens
Monograph No. 2
ECETOC, 1980

Medical Library Service
College of Physicians & Surgeons of B.C.
1383 W. 8th Ave.
Vancouver, B.C.
V6H 4C4

Risk Assessment of Occupational Chemical Carcinogens
Monograph No. 3
ECETOC, 1982

Assessment of Mutagenicity of Industrial and Plant Protection Chemicals
Monograph No. 9
ECETOC, 1987

OECD Guideline for the Testing of Chemicals
Introduction to the OECD guidelines on genetic toxicology testing and guidance on the selection and application of assays, Volume 2, 1993

1.6   Reproductive toxicology

No doubt anyone reading this is fully aware of the reproductive process – the process described to children as 'the birds and the bees'. From a more scientific standpoint, reproduction is the way in which new individuals of a given species are produced and genetic information passed on from parents to their offspring.

The reproductive process ranges from the development and maturation of both female and male reproductive systems, to successful mating and a resulting healthy, normal viable offspring. Our reproductive organs, which are referred to as gonads, contain special reproductive cells (gametes) and, as with any other organ or cell in our body, they are just as susceptible to chemical injury and therefore need to be taken into consideration when assessing any potential adverse effects from chemicals.

In the female the reproductive cell is known as the ovum or egg; in the male the reproductive cell is called the sperm. However, there are also other stages within the reproductive process that need to be looked at when investigating chemical and/or physical effects, as will be seen later.

Probably the most well-known incident highlighting the subject of reproductive toxicology hit the news back in the 1960s. It concerned the thalidomide drug which was given to pregnant women who suffered morning sickness. This had one appalling and unknown side-effect in that it gave rise to birth defects in the new-born when administered during the first twelve weeks of pregnancy (an important stage of pregnancy where the baby's organs develop, a process called organogenesis).

Although this example involves an administered drug as opposed to contact with potentially hazardous chemicals, the incident still highlights the importance of testing for the effects of a chemical over the whole reproductive cycle. Therefore, the aim of reproductive toxicity studies is to detect any immediate or delayed effects on mammalian reproduction arising from chemical exposure.

How chemicals cause harm is considered in Fig. 1.8. Chemicals at a number of different stages of reproduction, as is seen in this illustration, can affect the overall reproductive process. In the occupational setting, the effects of chemicals on reproduction is a relatively new area of concern and, unfortunately, there is not much information available for many of the chemicals that are used on a regular basis in industry.

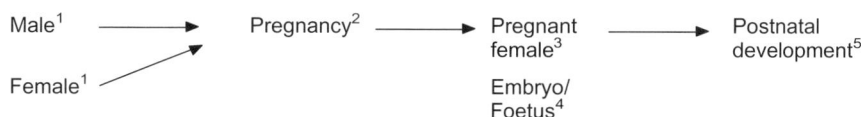

Male[1] ──────────▶ Pregnancy[2] ──────────▶ Pregnant ──────────▶ Postnatal
                                             female[3]   development[5]
         ↗
Female[1]   Embryo/
                                             Foetus[4]

Stage number

1. Chemicals can cause effects on the male and female reproductive cells/organs giving rise to fertility problems.

2. Could successfully conceive, that is, become pregnant but abortion occurs.

3. During pregnancy toxicity in the female could occur, with no persistent effects on the embryo or foetus.

4. Toxicity to the embryo/foetus, giving rise to death, structural malformations (teratogenicity) or other developmental abnormalities. No adverse effects seen with the mother.

5. Adverse chemical effects not seen until much later in life during the development of the individual.

Fig. 1.8   Different stages where chemicals could cause harm in the reproductive process.

1.6.1   *Reproductive toxicity testing*

This is split into two main areas for chemicals which are: (1) reproductive toxicology and (2) developmental toxicity.

As with other toxicological tests, the extent of testing required will depend on the anticipated exposure levels and results from initial studies. The route of exposure used in these studies is that anticipated in humans (usually oral).

1.6.1.1   *Reproductive toxicology*   This is concerned with the adverse effects of chemicals on the adult male and female reproductive process. These effects can include infertility, impotence, chromosome abnormalities and resulting birth defects, abortion and effects on lactation and impaired or delayed growth of the offspring.

Single- and multi-generation studies are used to investigate the potential of chemicals causing these types of reproductive disturbances over one and two reproductive generations, respectively. The rat or the mouse is usually the species of choice for such studies.

For a *single-generation study* the chemical is administered to both males and females prior to mating and continued in the pregnant female until the first generation of offspring are weaned (usually 21 days after birth), whereupon the test is stopped. Then all animals are sacrificed and examined for any reproduction disturbances. The results from this test can be used for subsequent testing.

The purpose of the *multi-generation study* is to assess the effect of the chemical on the reproductive process of the first generation and to detect any latent adverse effects that would not normally show up until after the first generation. A second generation is bred using the same methodology as above, whereupon all animals are sacrificed and examined in the same manner as previously. This study will detect any latent adverse effects not normally showing up until after the first generation. Perinatal and postnatal effects, including growth and development, are also examined in this test.

1.6.1.2 *Developmental toxicity* This is the adverse effect of chemicals on the normal developmental process of the offspring from the time it is inside the mother's womb (embryo/foetus) until postnatal development, where aspects such as behaviour and physical development are investigated. Although there are protective mechanisms and even a protective barrier in the form of the placenta present, these are not always enough to shield the developing embryo/foetus from chemical exposure via the mother. Chemicals can be toxic, lethal or cause birth defects in the developing embryo or foetus. Both the time of exposure, i.e. the point at which during pregnancy the chemical exposure occurred, along with the level of exposure (or dose), will play a critical role in determining the severity of the effect.

Birth defects occur as a result of different circumstances, the most well known being that due to exposing the developing embryo to a chemical called a teratogen. Teratogens cause physical or functional defects and in order for a teratogen to exert its effect, it must come into direct physical contact with the developing embryo and exposure must occur during the critical period of organogenesis. This is when major organ development occurs and as a result it is the period when the developing embryo is most susceptible to this type of chemical. More often than not, there is no associated maternal toxicity associated with such events.

However, not all birth defects are due to teratogenicity. They can also arise as a result of other circumstances, one being due to mutations occurring in the germ cells of the parents prior to mating, i.e. a congenital abnormality.

Developmental effects on the unborn are observed using teratogenicity tests most commonly on rats or rabbits. Once pregnancy has been established the test chemical is administered to the dam (pregnant female) during organogenesis, the period during which developmental abnormalities are most likely to occur. In the rat this is during the 6th to the 15th day and in the rabbit during the 6th to the 18th day. Prior to parturition (birth) the dams are killed and the developing foetuses are examined for any abnormalities. For pharmaceutical chemicals there is a separate test for perinatal and postnatal effects. However, for industrial chemicals, such effects are examined in the multi-generation study.

Further reading

Identification and Assessment of the Effects of Chemicals on Reproduction and Development (Reproductive Toxicology)
Monograph No. 5
ECETOC, 1983

Occupational Toxicology
Edited by Neill H. Stacey
Taylor & Francis, 1993

Basic Toxicology: Fundamentals, Target Organs and Risk Assessment, 3rd edn
Frank C. Lu
Taylor & Francis, 1996

General and Applied Toxicology, abridged edn
Edited by Bryan Ballantyne, Timothy Marrs & Paul Turner
Macmillan Press, 1995

OECD Guideline for Testing of Chemicals
Teratogenicity, No. 414

OECD Guideline for Testing of Chemicals
One-Generation Reproduction Toxicity Study, No. 415

OECD Guideline for Testing of Chemicals
Two-Generation Reproduction Toxicity Study, No. 416

2 Ecotoxicology

2.1 Introduction

For quite a few years now environmental issues have received considerable coverage from the media, which has resulted in an increase in public concern over the potential impact of chemicals on the environment. This concern has been reflected in industry as a whole, both in the increasing severity of regulations related to chemical use and in the heavy penalties for those who, intentionally or otherwise, violate them. Therefore, the aim of this section is to look at what is actually meant when we talk about the environment and what we need to know with respect to those chemicals we use on a daily basis. For our purposes the main area to be focused on is *ecotoxicology*, which can be defined as the study of the adverse effects of chemicals on the environment.

At this point the difference between ecotoxicology and toxicology should be clarified. Ecotoxicology is concerned with adverse effects caused by chemicals on large numbers of species, i.e. populations, communities, ecosystems, and it is a subject involving the direct study of the toxic effects on the organism of interest. In contrast, toxicological studies examine the adverse health effects of chemicals on the individual using animals or cell cultures as surrogates.

2.1.1 *The environment*

The *Collins English Dictionary* defines the environment as 'the external surroundings in which a plant or animal lives, which tend to influence its development and behaviour'. Taking a more detailed view, the environment is composed of both a living component, or biota, which includes animals, plants, humans, etc, and a non-living component, or abiota, such as air, water, soil, etc., each component interacting with the other in a variety of ways.

2.1.2 *Environmental compartments/media*

In the context of chemical exposure a common expression that is often used is 'environmental effects'. On its own this term can mean a lot of different things to different people and therefore it is really important to clearly specify in which context it is meant. In other words, 'what is being affected, or is at risk of being affected?' For example, is the effect on plants, animals or the aquatic environment, etc.

When attempting to be more specific with respect to where a particular chemical effect may occur, it is often useful to think of the environment as being made up of different compartments (or media). These are usually taken as being air, water, sediment/soil and biota (Fig. 2.1).

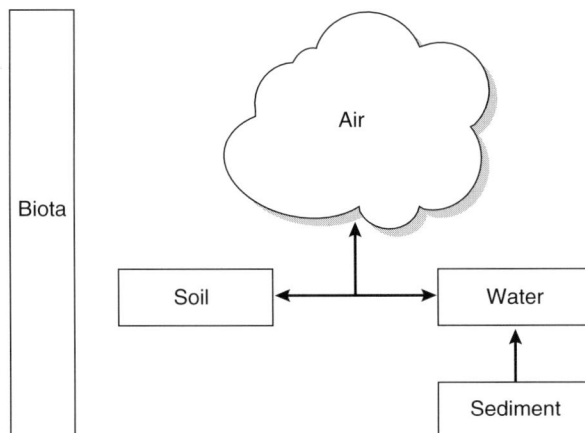

Fig. 2.1 Schematic representation of the environmental compartments.

2.1.3 *Environmental transport and fate*

Before a chemical can cause an effect it needs to get into one or more of the environmental compartments. However, as a result of natural phenomena such as winds and rainfall, or even the activities of humans, chemicals do not necessarily remain in the same compartment into which they were first released. Instead, they can be (geographically) transported within the same compartment long distances away from the original source of emission, or else they can transfer between different compartments, for example by the action of precipitation or evaporation.

The fact that there is constant change in this way can make life very difficult when trying to predict where chemicals will end up in the environment, i.e. their *environmental fate*. However, there are certain intrinsic properties of chemicals that can be used to help predict their potential movement, concentration and fate in the environment. These properties include volatility, water solubility, bioaccumulation potential and biodegradability. Although these intrinsic properties will influence the movement of chemicals and their fate in the environment, transport within the environment will, to a large extent, depend on the mobility of the medium into which the chemical was introduced. Therefore, chemicals present in air or water (compartments) will be transported further than those remaining within the soil or sediment. As a result, attention from environmental agencies world-wide is focused mainly on chemical pollution of both air and water media.

One well-known example of a problem resulting from chemical movement within the environment is that involving sulphur dioxide emissions from the United Kingdom being transported to countries in Scandinavia. This has caused the acidification of both water (lakes) and soil as a result of the gas being washed out of the air by water which then falls as 'acid rain'. Another example is pollution from landfill where leaching can cause hazardous chemicals originally bound to the soil to be transported into the groundwater or to evaporate and enter the air, thereby entering other environmental media.

2.1.4   *Ecotoxicological effects*

When investigating ecotoxicological effects, plants and animals in the environment can be studied as individuals, as part of a population, communities or even ecosystems.

The *individual*, such as a single bird, plant, fish, human, etc. is the basic level of ecological organisation. A *population* comprises groups of individuals of the same species existing within a defined geographical location. For example, populations of humans, daffodils, chestnut trees, etc. A *community* is essentially groups of different populations that coexist, again within some defined geographical location, and which interact with each other. An *ecosystem* comprises different communities along with the physical/chemical environment. Examples include forest ecosystems, aquatic ecosystems such as streams, lakes, etc. Another term, which is commonly used, is 'habitat'. This could be described as an organism's 'address' within the ecosystem.

Ecosystems are very robust and resilient to the natural forces of Nature, such as flooding, fire, drought, etc. However, the effects of chemical pollution can overwhelm their dynamic stability causing significant adverse effects. Probably the most well-known examples of chemical pollution are the deleterious effects of air pollution on rain forests and the effects of DDT (a pesticide which has been banned in many parts of the world) on species in the wild.

Within any ecosystem photosynthesising plants are primary producers and sit at the base of all *food chains*. These are sequences of organisms ascending from primary producers to first and secondary consumers at different trophic (or feeding) levels. In reality, as a result of the large number of interlinking food chains, a simple food chain as shown in Fig. 2.2 is seldom the outcome. Instead, a more complex *food web* is produced. As an example of a simple food chain in an aquatic ecosystem, a primary producer could be green plants (algae), a primary consumer could be invertebrates (crustaceans) and a secondary consumer would be fish. The tertiary consumer could be humans.

The concept of trophic levels is important because one of the aims of ecotoxicological testing is to assess the adverse effects of chemicals at these different levels. Within the aquatic environment, algae, invertebrates (usually daphnia) and verte-

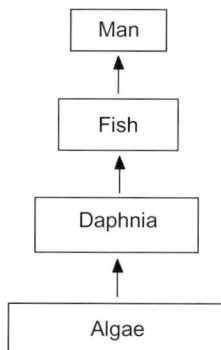

Fig. 2.2   A simple food chain, the arrows indicating the flow of food energy.

brates (fish) are taken as indicator species representative of their trophic level, the idea being that any effect seen at one trophic level could have the potential to affect either directly or indirectly other trophic levels.

2.1.5   *How, where and what is studied?*

Although it is possible to investigate the adverse effects of chemicals on all levels of biological organisation, it is most desirable to focus on the higher levels, ideally the ecosystems themselves, as it is believed that any adverse effects seen here will have a more significant impact on the environment as a whole. Although studies on these higher levels of organisation can be completed both in the field and even via the creation of laboratory ecosystems, they are in practice very expensive and time consuming to do and are not practical when trying to address routine regulatory requirements/risk assessment programmes. Therefore, studies are usually carried out using small groups of individuals, i.e. single species of biota with testing being done either in the laboratory or out in the field. Field studies attempt to study chemical effects under 'field conditions' which are similar (or identical) to those which would be expected in real life. However, these studies are usually not feasible because of the difficulty keeping track of the organisms being monitored and ensuring that the necessary measurements are continuous. Laboratory tests are more common and popular as they provide controlled conditions under which to study and assess the adverse effects of chemicals, as well as being economically viable and relatively simple to perform. Of course, questions are often raised as to their relevance when attempting to extrapolate to the 'real world'. This invariably is a question that is often posed when conducting laboratory studies both for ecotoxicological and toxicological effects. Consequently, the tests are carried out using species that are deemed to be most representative of their trophic level.

There are internationally recognised test methods available, mainly for aquatic environments, that use groups of species regarded as most representative of their trophic level. These species are vertebrates (fish), invertebrates (some kind of crustacean such as daphnia) and water plants (usually green algae). Such studies typically investigate factors including growth, death, development and reproductive capabilities. This is because it is believed that anything that may affect the size and the general health of populations is of importance.

Since aquatic studies are those involving standard test guidelines, which are recognised by international protection agencies, with results reported in safety data sheets, Section 2.2 will specifically relate to the aquatic environment.

2.2   Aquatic toxicity testing

The aquatic environment encompasses oceans, lakes, rivers, streams and ponds, etc. For the purposes of this chapter, attention will be given to the freshwater aquatic environment.

The purpose of aquatic toxicological testing is to detect any short-term/acute or long-term/chronic effects that may arise as a consequence of chemical exposure, taking particular consideration in those key indicators mentioned above, namely survival, growth, reproductive ability, etc. As will be understood after reading the following, many ecotoxicological studies simply involve observation and nothing more than that.

In contrast to toxicological studies, ecotoxicological tests use concentrations, e.g. mg/l which means milligram (test chemical) per litre (water).

2.2.1 Short-term (acute) and long-term (chronic) effects

The OECD guidelines define acute effects as 'lethal and sublethal effects observed after a short exposure period to the lifespan of the organism'. In the aquatic environment the end-point of acute toxicity studies is taken as being either death of or complete lack of movement (immobilisation) in the organism. This is because with some test species, such as fish, it is relatively easy to see whether or not they are dead and therefore to derive a lethal concentration value. However, with other species, such as invertebrates, it is very difficult to observe whether or not death has occurred because of their small size. Therefore, the end-point used is immobilisation and the measurement is usually an EC value, the effect concentration that produces immobilisation in a specified percentage of the exposed population. Standard tests are available for fish, daphnia and green algae.

As with mammalian toxicity studies, chronic studies use end-points other than lethality, i.e. observing sublethal effects, and from such studies it is possible to derive statistically NOEC (no observed adverse effect concentration) and LOEC (lowest observed adverse effect concentration) levels. The NOEC provides an estimation of the highest concentration that results in no observable adverse effects while the LOEC provides an estimate of the lowest concentration that produces an observable effect.

It is more difficult to assess chronic effects of chemicals in the aquatic environment than to assess their acute effects because of the limitations of carrying out long-term field or laboratory studies on aquatic organisms. There exist long-term standard studies for both fish and invertebrates but not algae. However, what is interesting to note is that the acute study, which is available for algae, is really a chronic test because many generations are exposed to the test chemical.

2.2.2 Test design

There are three test designs commonly used when studying the aquatic environment and the choice will depend upon the physico-chemical properties of the test chemical.

(1) *Static* – the test solution is not replaced and no flow occurs. This can be used if the compound is stable and non-volatile.
(2) *Semi-static* – the test solution is replaced at regular intervals. This is usually used if the test compound is not so stable.
(3) *Flow-through* – a constant flow rate and concentration of test solution is maintained at all times. This is used if the test compound of interest is volatile in nature.

2.2.3 *Test species*

As was mentioned above, studying the effects of chemicals in the aquatic environment usually requires three different groups of indicator species which are deemed to be representative of their trophic level, namely vertebrates, invertebrates and green plants. Some of the typical factors taken into consideration when selecting a particular species are summarised below:

- Availability
- Relevance
- Easily handled
- Sensitivity
- Database of background information

2.2.3.1 *Vertebrates* Fish (Fig. 2.3) are an important feature of many food chains and are the vertebrate chosen as being representative of their trophic level. There is a diverse range available for aquatic toxicity studies conveniently grouped according to whether the fish originate from a cold-water or warm-water environment. The choice of fish will depend upon practicalities such as availability and, of course, relevance to the area of interest. Thus, it is of far more interest to use a cold-water fish if the research is of specific relevance to countries with a cooler climate, such as those of Scandinavia.

Fig. 2.3   The most commonly used vertebrate in aquatic toxicity testing – the fish.

Probably the most common cold-water fish used in these kinds of tests is the rainbow trout, while the most commonly used warm-water fish are the sunfish, fathead minnow and zebra fish. Note that a problem readers may encounter is when research literature (correctly!) uses the Latin name of species. Table 2.1 provides a list of the most common fish used in ecotoxicological studies along with their 'formal' Latin names.

2.2.3.1.1 *Short-term effects: acute toxicity to fish*   The OECD 203 test method is commonly used where different groups of fish (seven per concentration) are exposed to different concentrations of the test chemical in their water over a period of 96 hours. At differing time intervals all mortalities are noted and the results are plotted as a concentration versus lethality graph. From this, the concentration at which half

Table 2.1 The fish most commonly used in ecotoxicological studies along with their 'formal' Latin name.

Common name	Latin name
Zebra fish	*Brachydanio rerio*
Fathead minnow	*Pimephales promelas*
Rainbow trout*	*Oncorhynchus mykiss** / (salmo gairdneri)*
Common carp	*Cyprinus carpio*
Red killifish	*Oryzias latipes*
Guppy	*Poecilia reticulata*
Blue gill	*Lepomis macrochirus*

* Rainbow trout is now referred to as *Oncorhynchus mykiss.*

of the fish die during the exposure period (the LC_{50} value) is derived for a given time point (usually 96 hours). From an ethical perspective, it is highly desirable to minimise the numbers of fish used in these kinds of studies. Therefore, there is also a limit test available which uses a group of seven fish exposed to a concentration level of $100 \, \text{mg} \, l^{-1}$ of test substance. (Note that the US EPA test method requires 10 fish per concentration level and studies are completed in duplicate.) If there are no mortalities then the LC_{50} value is taken as being $> 100 \, \text{mg} \, l^{-1}$. Should mortalities occur, then the full study needs to be carried out.

2.2.3.1.2 Long-term effects: chronic toxicity to fish There are studies that investigate the adverse effects of chemicals arising from exposure to chemicals over a longer period of time. These are:

(1) Prolonged 28-day exposure. Like the acute toxicity test, this study uses survival as an end-point along with effects on growth, thereby producing an LC_{50} and EC_{50} value, respectively.
(2) Fish early-life stage test. Here, newly fertilised eggs are placed in the test system and any adverse effects on hatching, and sub-lethal effects such as growth and survival are recorded. The fathead minnow is usually the preferred species because it has a relatively rapid egg-hatching period.
(3) Life cycle studies. This type of study investigates the adverse effects of chemicals on the whole life cycle (egg to egg) and in some cases may take up to two years to perform. In particular, focus is given to hatching success, development and growth and fertility and egg viability.

2.2.3.2 Invertebrates Daphnids (water fleas; see Fig. 2.4) are an important part of fresh water ecosystems and comprise an important source of food for many fish. They are herbivorous filter feeders in nature, which means that they come into close contact with the aquatic environment.

From a laboratory testing perspective different types are used – *Daphnia magna, Daphnia pulex* and *Ceriodaphnia dubia* are all examples. Each is conveniently cultured, representative of its trophic level and relatively sensitive to the effects of chemicals. Other invertebrate species (such as molluscs) can be used in ecotoxicity testing, but daphnids are probably most commonly seen in freshwater studies. Both

Fig. 2.4 Daphnia are commonly used invertebrates in aquatic toxicity testing.

acute tests and sub-lethal effects, such as reproductive disturbances, can be investigated.

2.2.3.2.1 *Short-term effects: acute toxicity testing*   As mentioned above, acute toxicity tests involving invertebrates measure immobilisation. Therefore, no LC_{50} value is derived. Instead an EC_{50} value, which is the 'effect-concentration' where 50% of the test population is immobilised within the specified time period, is measured. The acute toxicity test as described within the OECD guidelines (OECD 202, Part I) uses a range of different test chemical concentrations in order to investigate their effects on the capability of young daphnia to swim over an exposure period of up to 48 hours, a period that allows them to moult at least once during which time they are very sensitive to chemical effects. The results are plotted and the concentration at which 50% of the test population is no longer swimming, the EC_{50}, at 48 hours is then derived.

2.2.3.2.2 *Long-term studies*   The test listed in the OECD guidelines is the Daphnia reproduction study (OECD 202, Part II). In this test, lower concentrations (compared to the acute study) of test chemical are used over a longer period of time, usually 21 days, so as to assess effects on reproduction, growth and survival.

It is a sensitive test and a decline in the reproduction rate will quickly be seen if the daphnia are affected by the test chemical. Both the LOEC (the concentration which produces the first significant change in numbers compared to the control) and NOEC can be derived from the results. Likewise, once the shape of the concentration–response plot has been established, it is possible to derive any EC value (or LC value), for instance EC_{10}, EC_{100}, etc. (see Fig. 2.5).

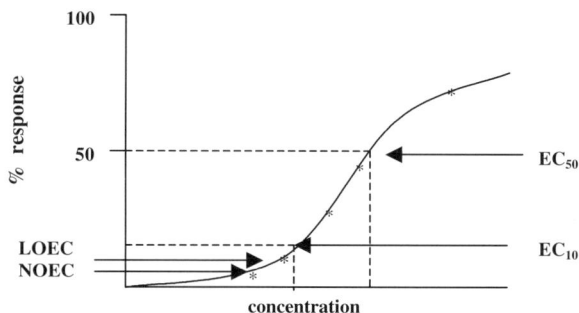

Fig. 2.5 Graph showing a typical concentration–response curve.

2.2.4   *Aquatic plants*

Within the aquatic environment plants produce oxygen and constitute an important trophic level because they form the base of many food chains (primary producer). In aquatic studies, green algae are used as a surrogate for aquatic plants in tests using fresh water. The species typically used are *Selenastrum capricornutum, Scenedesmus subspicatus,* and also (less commonly) *Chlorella vulgaris.* Since algae have a short life cycle, such species respond quite quickly to any changes within their surroundings and the effects of chemicals on algae will manifest themselves either as an increase in growth (algae bloom) or inhibition of growth, both of which are undesirable.

The algae growth inhibition test (OECD 201) is the one most commonly reported in safety data sheets for freshwater systems and, since it involves the exposure of many generations of algae during the test period, should really be defined as being a measurement of chronic toxicity. The test involves the measurement of algae when exposed to different concentrations of test chemical over a period of 72 hours. The results are plotted and the EC_{50} value is determined. In this case, the EC_{50} is the concentration at which there is a 50% reduction in growth. (Note that a principal difference in the test methods of the OECD and the EPA is that the OECD guidelines use a test duration of 72 hours whereas the EPA guidelines use an exposure period of 96 or 120 hours.) Occasionally an IC_{50} value will be seen in the research literature. This denotes the concentration at which there was 50% growth inhibition. The value obtained and method used are identical to those described above where an EC_{50} value is given. It is simply a different nomenclature.

2.2.5   *Summary*

Aquatic ecotoxicological tests are carried out for a number of reasons. These include product notification purposes, the granting of permits and discharge consents, and environmental risk assessments. Overall, it is very important to learn more about the environment and how chemicals can influence its 'well-being'. This knowledge goes some way to ensuring its conservation, and since humans are part of the environment too, any damage we cause either knowingly or in ignorance, will ultimately impact on our own well-being.

Further reading

Fundamentals of Aquatic Toxicology: Effects, Environmental Fate, and Risk Assessment, 2nd edn
Edited by Gary M. Rand
Taylor & Francis, 1995

Chemical Ecotoxicology
Jaakko Paasivirta
Lewis Publishers, 1991

Pollution: Causes, Effects and Control, 2nd edn
Edited by R. M. Harrison
Royal Society of Chemistry, 1990

OECD Guideline for Testing of Chemicals
Algae, Growth Inhibition Test, No. 201

OECD Guideline for Testing of Chemicals
Daphnia spp., Acute Immobilisation Test and Reproduction, No. 202

OECD Guideline for Testing of Chemicals
Fish, Acute Toxicity Test, No. 203

OECD Guideline for Testing of Chemicals
Fish, Prolonged Toxicity Test: 14-day Study, Test No. 204

OECD Guideline for Testing of Chemicals
Fish, Early-Life Stage Toxicity Test, No. 210

2.3   Chemical persistence and bioaccumulation

2.3.1   *Introduction*

If a chemical is said to be degradable it is meant that it can be broken down into a simpler chemical structure(s), which can either lead to a chemical being mineralised, i.e. totally degraded to carbon dioxide, simple salts and water, or to being partially broken down, which can cause more problems if the resulting residue is more harmful to the environment than the parent compound. A chemical that is only slowly or not easily degraded is called *persistent*. Degradation occurs by three main processes (Fig. 2.6): biologically; photolytically (using sunlight); and chemically, either in the presence of air (aerobic) or in its absence (anaerobic).

Biological degradation or *biodegradation* is a biotic process, that is it involves living organisms, whereas both photolytic and chemical degradation are abiotic processes. As was outlined in Section 2.2, the protection of both air and aqueous media is what most environmental agencies prioritise highly. Therefore, for our purposes, the following sections will deal with chemicals in the aquatic environment.

2.3.1.1   *The aquatic environment*   In the aquatic environment the two most important mechanisms for the degradation of chemicals is either by a chemical process called hydrolysis or by biodegradation. *Hydrolysis*, as the name implies, involves the splitting of the compound by water while *biodegradation* is the degradation of chemicals by biological means, in other words by the action of micro-organisms, or 'bugs'!

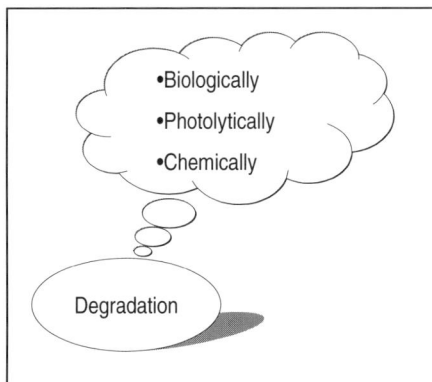

Fig. 2.6 The three main processes by which degradation can occur.

2.3.2 *Biodegradation*

During biodegradation micro-organisms feed off the chemical substrate, breaking it down and using it in order to grow. This process is known as *respiration* and it involves the consumption of oxygen. What should be noted is that only organic compounds can be biodegraded and the majority of these will eventually break down in the environment, though the actual rate of degradation varies as some chemicals will degrade more rapidly than others. (Inorganic chemicals such as sodium silicate cannot be biodegraded, but they may be degraded by some other means.)

The degree to which a chemical degrades will determine just how much of the chemical will remain in the environment after a specified period of time. However, just because it will degrade in the environment does not necessarily mean that this is the end of the story. Assuming that the chemical in question is not completely mineralised its products of degradation, or metabolites, need also to be considered because in some cases these can be more harmful to the environment than the original parent compound. A very good example of this situation involves nonyl phenol ethoxylate the degradation product of which, nonyl phenol, gives rise to problems in the environment.

2.3.3 *Tests for biodegradability in the aquatic environment*

Because the real environment is variable and as a consequence hard to reproduce accurately, it is very difficult to obtain an accurate picture of the true biodegradability of a chemical under laboratory conditions. However, laboratory tests are the way in which biodegradation is assessed and a tiered approach is usually adopted so as to minimise costs without losing sight of the ultimate objective, namely the protection of the environment. The approach is shown in Fig. 2.7.

2.3.3.1 *Ready biodegradability* The first test, and the one very often seen in literature and requested by regulators, is the *ready biodegradability test*. This provides very limited opportunity for both biodegradation and adaptation of the microbes to

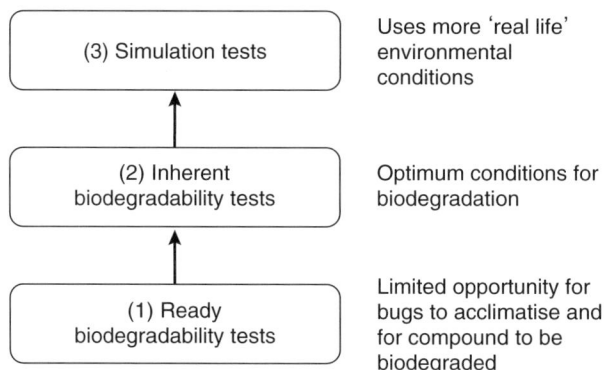

Fig. 2.7   Schematic representation of the approach used for testing biodegradability.

the test substance and, as a result, is a tough test for a chemical 'to pass'. Therefore, the assumption is made that any chemical passing this test will readily biodegrade in the aquatic environment without any problem.

The OECD guidelines provide a number of tests which can be used to assess ready biodegradability and the biodegradation levels needed in order to qualify as being readily biodegradable. The choice of the test is usually based upon the intrinsic properties of the test chemical, such as solubility and adsorption characteristics, etc. All the tests last for a period of 28 days and use a non-specific method of analysis in order to follow the biodegradation process: some use dissolved organic carbon (DOC) whereas others use CO_2 evolution. The analytical methods used in the tests have been indicated in Table 2.2.

Table 2.2   OECD tests available for ready biodegradability testing.

Test methods available	Pass level (%)	Analytical method
OECD 301A	70	DOC
OECD 301B	60	CO_2 evolution
OECD 301C*	60	O_2 consumption
OECD 301D	60	Dissolved oxygen
OECD 301E	70	DOC
OECD 301F	60	O_2 consumption

2.3.3.1.1   *Ten-day window – what is this?*   Anyone who has read a ready biodegradability test report will have seen reference made to the '10-day window' and have probably wondered what exactly it is. Well, in order for a chemical to pass the test for ready biodegradability, it is necessary for it to have attained the pass level of degradation within a time window of 10 days. This 10-day window starts as soon as 10% biodegradability is achieved within the test. Therefore, if a result shows that a product has achieved 76% biodegradability within 28 days and it only achieved 45% within the 10-day window, then the product cannot be deemed as being readily biodegradable. The 10-day window concept is used in order to prevent substances passing the test that are only slowly degraded, but have a short period of acclimatisation.

Note that there is one ready biodegradability test for which the 10-day window does not apply – this is follows the OECD 301C* guideline, the MITI test.

2.3.3.2 *Inherent biodegradability* Just because a chemical is not readily biodegradable does not necessarily mean bad news in the sense that it may persist in the environment. Before any such judgement can be passed, it is necessary to carry out further tests in order to assess whether or not there is any potential for biodegradation. Going to the next stage as shown in Fig. 2.7 and carrying out a test for inherent biodegradability does this.

Two of the most common tests for determining inherent biodegradability are the Zahn–Wellins test (OECD 302B) and the modified SCAS test (OECD 302A). Both of these tests use conditions that are optimal for biodegradation, i.e. they permit a longer exposure period (not restricted to 28 days) together with a higher concentration of micro-organisms. Should the results indicate that more than 20% biodegradation has occurred, then the test substance can be classed as *inherently biodegradable*. Should greater than 70% biodegradation occur then the test substance could be classed as *ultimately biodegradable*.

However, with these tests, it can never be assumed that out in the 'real' environment biodegradation will definitely occur and quickly. If the biodegradation is very poor in this test (< 20%) then in the context of estimating environmental concentrations of chemical substances in different compartments, non-biodegradability (persistence) would have to be assumed. However, it must be emphasised that just because a chemical does not biodegrade in this or other biodegradation tests, there are still other ways in which it could degrade in the aquatic environment, e.g. via hydrolysis, and this should also be taken into consideration.

2.3.3.3 *Simulation tests* These tests use conditions that are close to those expected in the 'real' environment. However, owing to their complexity and expensive set-up costs, such tests are usually restricted to biodegradation research projects simulating different aquatic ecosystems such as those found in lakes and rivers.

2.3.4 *BOD and COD tests*

As was mentioned above, all organic material can be decomposed by micro-organisms. Under aerobic conditions the micro-organisms use oxygen as they break down this material. Unfortunately, in doing so, there is a risk that a considerable proportion of the dissolved oxygen present in the aqueous environment will be consumed to such an extent that many other aquatic organisms that depend on this gas may not survive.

The *biological oxygen demand*, or BOD, can be used to measure the amount of oxygen that is consumed by micro-organisms under aerobic conditions during a specified time period. This period can be set at five days or more, and 20 or 28-day BOD values are not uncommon. Overall, the BOD value is a useful indirect indicator of just how polluted is a water source such as a river or effluent. For example, water that is highly contaminated with organic material will have a high BOD value whereas drinking water will (hopefully!) have a very low BOD value. If an organic chemical such as cationic starch were to be accidentally released into a lake, then the existing BOD value would increase although by what extent would of course depend on the amount released, and the characteristics of the receiving water.

BOD levels can also increase, not as a result of the chemical being released *per se*, but due to its indirect effects on the environment. Eutrophication arising from the release of nitrates and phosphates into aquatic systems has caused problems in the form of algae bloom (excess algal growth). When the algae begin to die, the concentration of organic material in the water increases. This causes an increase in micro-organisms as they feed off this material and decompose it, diminishing the levels of dissolved oxygen and threatening other aquatic species which depend on the oxygen.

Chemical oxygen demand, or COD, is the amount of oxygen needed to chemically oxidise completely any water-borne compounds (not just the organic variety) and is a measure that approximates to the theoretical oxygen demand (ThOD), which is attained upon complete oxidation of the chemical.

COD and BOD values are just two of a number of parameters that are requested and used by local authorities when characterising the discharge of effluents to receiving waters.

Since the restrictions imposed with respect to permitted discharge limits are tough and penalties for infringement are great, it is important and useful to know which process or chemical is a large contributor to these figures. In this way it is then possible to reduce the charge to the effluent if needed.

Biological oxygen demand alone can give an indication of biodegradability, in that a chemical with a high BOD value can be assumed to have the potential to biodegrade in the environment. However, COD values give no indication whatsoever of any potential biodegradability, unless used in conjunction with the BOD value (see Table 2.3). The ratio of these two values can be used as a 'screening test' in order to ascertain whether or not a chemical will be readily biodegradable:

BOD/COD × 100 = Biodegradation (%)

It must be noted that this is only an approximation (but a useful one!) and as such should be used with care. For notification purposes, regulators will almost certainly require a complete ready biodegradability test.

Table 2.3 The ratios of BOD and COD that can be used to assess ready biodegradability with the percentage pass levels needed. Note that the ThOD values may be substituted for COD values.

Ratio	Assessment level
BOD_5/COD	> 50% biodegradation
BOD_{28}/COD	> 60% biodegradation

2.4   Bioaccumulation

It is commonly known that organisms living in aquatic environments have the ability
to absorb chemicals from their immediate surroundings be it via their gills, outer
protective layers or by direct ingestion of contaminated water or food. However, the
levels of chemical contamination we are considering here can be very low indeed.
Once absorbed, these chemicals can be metabolised, excreted, or otherwise, they can
be stored within the fatty tissue layers of the organism in a process known as
bioaccumulation. Dioxins are examples of chemicals known to have the potential to
bioaccumulate. Not all chemicals will bioaccumulate – in fact those chemicals that
demonstrate this tendency are usually organic, non-ionised and very poorly soluble or
insoluble in water. In general it is lipophilicity and water solubility that will determine
just how strongly the tendency of a chemical will be to bioaccumulate. (Lipophilic
means 'fat loving'.)

The process of bioaccumulation can occur either directly, in which case chemicals
are taken up by the organism immediately from its surroundings, or indirectly where
chemical uptake usually proceeds via the food chain. In the aquatic environment,
direct bioaccumulation of chemicals in living organisms is a common process. The
reason is that aquatic organisms, such as fish, are invariably in close contact with their
immediate surroundings and therefore any chemical contaminants present in the
water. Under these circumstances, once the level of intake exceeds the level of
excretion, bioaccumulation could be expected and eventually a steady state situation
would occur where the rate of uptake would equal to the rate of excretion.

Of concern is the fact that the level of chemical contamination can be significantly
higher in the organism than in its aquatic surroundings, as is the case with the
pesticide DDT. This can be measured and expressed as *the bioconcentration factor*,
which is the ratio of the concentration of chemical in the fatty layers of the organism
to that found in its surroundings. For terrestrial organisms, including humans, uptake
of chemical contaminants occurs mainly in an indirect manner, usually via the food
chain. This is a cause of great concern because once chemicals enter the food chain
there is an increased risk of higher trophic levels being exposed to high levels of
contamination as a result of *biomagnification*. The best known example of bioaccumulation and resultant biomagnification is with DDT.

2.4.1   *Testing for bioaccumulation*

For non-ionic chemicals there is a simple test that can be used in order to determine
their bioaccumulation potential. This is called the octanol–water partition coefficient
test and is based on the fact that both water solubility and lipophilicity determine the
potential of an organism to bioaccumulate a chemical. It uses a 'surrogate organism'
(which is not actually a living organism at all but a special piece of glassware) that
contains two phases – water and n-octanol. The n-octanol is an organic compound
that is used to mimic the fatty layers of aquatic organisms and the measurement
involves observing the ability of the test chemical to separate into these layers from
which the partition coefficient (denoted P_{ow}) can be derived. The resultant value can
then lead to a prediction of bioaccumulation, as summarised in Table 2.4.

Table 2.4 Interpretation of bioaccumulation potential from the partition coefficient test.

Coefficient value	Potential
$\log P_{ow} > 3$	Bioaccumulation potential is high
$\log P_{ow} < 3$	Bioaccumulation potential is low

2.4.2 *Bioconcentration study*

For some chemicals it is not possible to use the n-octanol–water test in order to determine the potential of a chemical to bioaccumulate. Such chemicals include metallo-organic compounds and inorganic compounds. For these chemical types, a fish study must be carried out. This study must also be performed on those organic chemicals where a $\log P_{ow}$ value equal to or greater than three has been obtained in the n-octanol–water test. The fish study usually involves a radio-labelling technique.

Fish are the chosen test species and are exposed to the test chemical over a period of 28 days whereupon they are placed in clean water for 14 days in order to provide an environment for the test chemical to appear should it be leached out of the fish. After the 14 days the ratio of concentrations of test chemical found in the fish to that found in the aqueous medium is calculated and a bioconcentration (B_{CF}) value is derived. Table 2.5 indicates how such values are interpreted.

Table 2.5 Interpretation of bioaccumulation results.

Bioaccumulation	Potential
$B_{CF} > 100$	Bioaccumulation potential is high
$B_{CF} < 100$	Bioaccumulation potential is low

2.4.3 *A commonly asked question: 'How do I know whether or not a chemical will upset my waste water treatment process?'*

The Activated Sludge, Respiration Inhibition Test (OECD 209) can be used when attempting to determine whether or not a chemical will interfere with the normal waste water treatment processes that most pulp and paper mills have on site.

The test, which lasts for three hours, uses activated sewage sludge. During the test procedure the inhibitory effects of different concentrations of chemical on microbial respiration are assessed by the measurement of their oxygen consumption rate. The value obtained from such a test is the EC_{50}, the concentration at which the respiration rate of the microbes decreases by 50% (compared to the control).

As is pointed out in the OECD guidelines, the EC_{50} value obtained in such a test only gives an indication of any possible toxicity to the micro-organisms present in the activated sewage sludge. It is usually taken that a chemical giving an EC_{50} value greater than 100 mg l^{-1} would not present a problem although one with an EC_{50} value that was less than 10 mg l^{-1} would be of concern. This test is itself not used for any classification purpose under European guidelines, but instead used in the risk assessment process.

2.5 Summary

(1) Chemicals which are only slowly or not easily degraded are *persistent*.
(2) There are three main degradation processes: biological, chemical and photolytic.
(3) The ratio of biological oxygen demand (BOD) and chemical oxygen demand (COD) can be used to ascertain whether or not a chemical would be biodegradable
(4) Lipophilic chemicals can be stored in the tissues of organisms and then give rise to cumulative toxicity.
(5) Bioconcentration is the process by which chemicals enter the tissues of the organisms from aquatic environments and are accumulated.
(6) Bioaccumulation is where chemicals are bioconcentrated and also accumulated as a result of the consumption of food.
(7) Biomagnification is where the concentration of accumulated chemicals increases as the chemical passes up the food chain.

Further reading

Fundamentals of Aquatic Toxicology: Effects, Environmental Fate, and Risk Assessment, 2nd edn
Edited by Gary M. Rand
Taylor & Francis, 1995

Chemical Ecotoxicology
Jaakko Paasivirta
Lewis Publishers, 1991

Pollution: Causes, Effects and Control, 2nd edn
Edited by R. M. Harrison
Royal Society of Chemistry, 1990

OECD Guideline for Testing of Chemicals
Ready Biodegradability, No. 301

OECD Guideline for Testing of Chemicals
Inherent Biodegradability, Modified SCAS Test, No. 302A

OECD Guideline for Testing of Chemicals
Zahn–Wellins/EMPA Test, No. 302B

OECD Guideline for Testing of Chemicals
Bioaccumulation Tests in Fish, Nos. 305A–305E

Medical Library Service
College of Physicians & Surgeons of B.C.
1383 W. 8th Ave.
Vancouver, B.C.
V6H 4C4

3 Classification and labelling of chemicals according to their hazardous nature

3.1 Introduction

The aim of classification and labelling of chemicals is primarily to ensure that the user is made fully aware of any potential hazards that may be associated with their use. Most countries have their own system for the classification and labelling of hazardous chemicals, the information being conveyed to the user in the form of a label and Material Safety Data Sheet (MSDS).

Since it is becoming increasingly common to buy chemicals from different parts of the world, it is useful to be aware of the legislation and criteria used by different countries for the classification and labelling of hazardous chemicals. Also, an awareness of the rationale behind the classification and labelling system provides the knowledge with which it is necessary to assess the information provided by suppliers or the corresponding MSDS and thereby determine whether or not the classification and information as a whole are correct. The current situation with respect to the classification and labelling systems for Europe, USA and Canada have been summarised and outlined in the following sections.

These summaries are not meant to replace any official documents, however, and readers are therefore strongly urged to obtain copies of relevant labels or MSDS for their own use of unknown chemicals in the workplace. Instead, the aim of this chapter is simply to highlight the similarities and differences in classification, labelling, and the criteria used in Europe, the United States and Canada.

One last comment. Although all these countries list criteria for physico-chemical hazards, such as oxidising properties, flammability, etc., these will not be mentioned here. Rather, the focus will be on both health and environmental aspects of the hazard. The reader is referred to the relevant official documents listed in the 'Further reading' section for more information on other hazards.

3.2 Europe

Within the European Union (EU) there exist two directives covering the classification, labelling and packaging of hazardous (or dangerous) chemicals. These are the Dangerous Substances Directive 67/548/EEC and the Preparations Directive 88/379/EEC. Both directives are implemented at the national level with each member state having incorporated them into their national legislation.

3.3 The Dangerous Substances Directive 67/548/EEC

The Dangerous Substances Directive (DSD) covers the classification, labelling and packaging of dangerous substances (household and industrial) supplied to or within

the EU that are not regulated by other product-specific legislation such as the Cosmetics Directive etc. Within the DSD a substance is defined as 'chemical elements and
their compounds in the natural state or obtained by any production process...'. For
the full definition the reader is referred to Directive 67/548/EEC.

A substance is considered to be dangerous, and therefore regulated by the DSD, if
it is classified in any of the categories shown in Table 3.1. Hazard symbols, which are
used in association with Table 3.1, are shown in Fig. 3.1. These symbols are always
orange and black and below them is written an indication of the corresponding
danger pertaining to the symbol. However, in one instance, flammable substances, a
symbol is not allocated.

The process of classification involves the placing of a chemical substance into a
particular category of danger (see Table 3.1) by identifying the hazard based on
criteria stipulated in the DSD, followed by the assignment of applicable risk
phrase(s).

Classifications cannot be unilaterally removed by a supplier of a chemical on the
grounds that the intended use of this chemical would be such that the probability of
hazardous exposure is zero. This is because classifications are based on the intrinsic
hazardous properties of the chemical substance and therefore, the user should take
care not to confuse 'hazard' with 'risk'.

Table 3.1 The different categories of danger for chemical substances.

Category of danger	Indication of danger	Symbol letter
Health related		
Very toxic	Very toxic	T+
Toxic	Toxic	T
Harmful	Harmful	Xn
Corrosive	Corrosive	C
Irritant	Irritant	Xi
Sensitising:	Harmful	Xn
	Irritant	Xi
Carcinogenic:		
Carc. Cat. (1&2)	Toxic	T
Carc. Cat. (3)	Harmful	Xn
Mutagenic:		
Mut. Cat. (1&2)	Toxic	T
Mut. Cat. (3)	Harmful	Xn
Toxic to reproduction:		
Repro. Cat. (1&2)	Toxic	T
Repro. Cat. (3)	Harmful	Xn
Physico-chemically related		
Extremely flammable	Extremely flammable	F+
Highly flammable	Highly flammable	F
Flammable	–	–
Explosive	Explosive	E
Oxidising	Oxidising	O
Environmentally related		
Dangerous for the environment	Dangerous for the environment	N or/and R52, R53, R59

Source: Dangerous Substances Directive and CHIP (1994).

HARMFUL/IRRITANT

CORROSIVE

(VERY) TOXIC

OXIDISING

HIGHLY OR EXTREMELY
FLAMMABLE

EXPLOSIVE

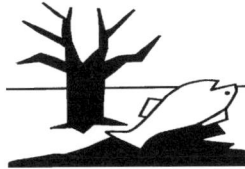

DANGEROUS FOR
THE ENVIRONMENT

Fig. 3.1 Common hazard symbols as part of the Dangerous Substances Directive 67/548/EEC.

3.3.1 *Risk phrases*

Risk phrases (or R-phrases) are used to describe a hazard. Numerous phrases are available and in some instances more than one risk phrase can be applied to a given category of danger. When considering combinations, however, the following should be noted:

(1) If two risk phrases are written in the following manner, R45-23, this means that each phrase is relevant, that is *R45 May cause cancer* and *R23 Toxic by inhalation*. The phrase as shown does not mean R45 to R23 inclusive!

(2) Risk phrases written together in the manner R20/21 means that the phrases are combined into a single statement. Thus, *R20/21 Harmful by inhalation and in contact with skin*.

3.3.2 *Safety phrases*

These phrases provide advice on the precautions to be taken in order to avoid exposure to a hazardous chemical and the measures to take should exposure actually occur. For example, *S36 Wear suitable protective clothing*, *S18 Handle and open container with care*.

3.3.3 *Labelling*

Labelling is the method by which information is conveyed to the user of any intrinsically hazardous chemical substance. The actual label is designed to provide information related to the hazard in the simplest manner possible using symbols shown in Fig. 3.1. The hazard is usually indicated together with applicable risk and safety phrases. Although the symbol and classification given on the label will indicate the hazard of principal importance, any other hazards should also be described by appropriate risk phrases. When more than one danger symbol could be assigned, the following order is usually adopted (see Table 3.1 for symbol meanings): symbol T takes precedence over C, which takes precedence over Xn, which takes precedence over Xi (i.e. T > C > Xn > Xi). For example, a chemical substance could be classified as harmful, but it could also be a skin sensitiser. The symbol and classification would reflect the fact that it is harmful, but the risk phrase *R43 may cause skin sensitisation* should also be given.

A label should contain the name, address and telephone number of the supplier, the product name, an indication of the danger along with a symbol (if applicable) and the relevant risk and safety phrases. Last but not least, the classification, symbol, and risk and safety phrases given on the label should be identical to those of the corresponding MSDS. A label is not a replacement for an MSDS, or vice versa.

3.3.4 *Classification criteria*

Within the DSD, criteria are given relating to how a chemical substance should be classified. The criteria are usually based on the results obtained from specific animal or, in some cases, validated *in vitro* studies. The test methods recommended, which are listed in Annex V of Directive 67/548/EEC, are based on OECD guidelines. Occasionally, adequate evidence exists showing that the toxic effects of chemical substances on humans are either different or likely to be different from those effects found in the animal test studies. In such instances the classification is based on the human evidence.

By agreement of countries within the European Union, Annex I of the Dangerous Substances Directive 67/548/EEC has a list of chemical substances that have been pre-classified and assigned risk and safety phrases that must be followed. As an example, sodium hydroxide is listed in Annex I and has been pre-assigned with both a symbol and risk/safety phrases. The pre-classification details are shown in Table 3.2. In cases where a chemical substance is not listed in Annex I of the DSD, suppliers are required to use the criteria given in Annex VI Part IID and 'provisionally classify' the chemical substance themselves.

Table 3.2 Annex I classification for sodium hydroxide. Note how the classification is based on the different concentration levels.

Sodium hydroxide	Classification
Concentration $\geq 5\%$	C; R35
$2\% \leq$ concentration $< 5\%$	C; R34
$0.5\% \leq$ concentration $< 2\%$	Xi; R36/38

The following sections will provide summaries of the criteria used for the classification of chemical substances. For the purpose of this book, only health and environmental effects will be summarised. For other hazardous effects the reader should refer to the Dangerous Substances Directive and its amendments. Note that it is not the intention that these summaries are used as a replacement for the Dangerous Substances Directive. Readers are strongly urged to obtain a copy of the Directive and its amendments if their intention is to classify and label substances for supply into or within Europe.

3.3.4.1 Criteria used for very toxic, toxic and harmful chemicals

For adverse systemic effects arising from acute toxicity, repeated dose or non-lethal, irreversible effects, there are three categories to which a chemical could be assigned. These are 'Very Toxic', 'Toxic' and 'Harmful', respectively.

Table 3.3 shows details of the criteria for these categories, obtained from animal test results where the classification was based on *acute lethal effects*. (Note that for the acute oral route other criteria are specified in the Directive should the fixed-dose method be used.)

Table 3.3 The criteria for classification based on acute lethal effects. All of the categories have a corresponding R-phrase.

Category	Oral, rat (mg kg^{-1})	Dermal, rabbit or rat (mg kg^{-1})	Inhalation, rat, 4 h (mg l^{-1})
Very Toxic	$LD_{50} \leq 25$	$LD_{50} \leq 50$	$LD_{50} \leq 0.50$ (gases and vapours)
Toxic	$25 < LD_{50} \leq 200$	$50 < LD_{50} \leq 400$	$0.5 < LD_{50} \leq 2$ (gases and vapours)
Harmful	$200 < LD_{50} \leq 2000$	$400 < LD_{50} \leq 2000$	$2 < LD_{50} \leq 20$ (gases and vapours)

The categories 'Very Toxic', 'Toxic' and 'Harmful' can also be assigned to chemical substances based on *non-lethal, irreversible effects* arising from a single exposure to the chemical. In this situation the assignment of one of these categories would be based on similar dose levels as those used for acute lethality. However, this does not include those chemicals that are carcinogens, mutagens or any giving rise to reproductive effects. These chemicals have specific criteria, which are described in Section 3.3.4.1.1.

The categories 'Toxic' and 'Harmful' can also be assigned to chemicals that cause *severe effects after repeated or prolonged exposure*. These assignments would be based on findings from subacute, subchronic or chronic studies and the criteria adopted are given in the Dangerous Substances Directive.

3.3.4.1.1 *Specific effects on human health: carcinogens, mutagens and substances toxic for reproduction*   For the health effects associated with each of these substances, there are three different categories together with assigned risk phrases, as shown in Table 3.4.

3.3.4.2 *Irritancy*   Classifications associated with irritancy can be based on either animal studies or practical observations in humans. Chemical substances irritating to skin or eyes or the respiratory tract as per the criteria given in Table 3.5 are classified as 'Irritant' and assigned the symbol Xi. The risk phrases used are also shown in Table 3.5.

3.3.4.3 *Corrosivity*   A substance is classified as 'Corrosive' to living tissues if it causes full thickness destruction of the tissue in at least one animal during the test for irritancy as given in Annex V of Directive 67/548/EEC. (Details of this test were given in Section 1.3.) For corrosive effects, a chemical substance can also be classified based on validated *in vitro* studies. Corrosive chemicals are assigned the symbol C.

For ethical reasons, chemicals or preparations are not required to be tested if they have a known pH of 2 or less or a pH greater than or equal to 11.5. A chemical's acid or alkaline reserve should also be taken into account. In these instances, such chemicals would automatically be classified as 'Corrosive'. There are two risk phrases associated with the 'Corrosive' classification and these are used when certain criteria, as shown in Table 3.6, are met.

If a chemical substance is classified as 'Corrosive' then it is assumed by implication that the same effects would occur if the chemical came into contact with the eyes. Therefore, a classification of corrosive (C) is applicable to both skin and eyes (even though the test is done on the skin only). Note that a chemical labelled corrosive will automatically be classified as corrosive for transport, even though it might not necessarily be corrosive to any metal container in which it is carried.

3.3.4.4 *Sensitising*

3.3.4.4.1 *Sensitising by skin contact*   The criteria for this effect are based either on knowledge that the substance can cause a sensitisation reaction in a substantial number of people or should positive results be obtained in appropriate animal studies.

Chemical substances that have been demonstrated to cause sensitisation by skin contact are classified as 'Sensitising' and assigned the symbol Xi, Irritant. The following risk phrase would then be used: *R43: May cause sensitisation by skin contact.*

As was mentioned in Section 1.3, there are two main tests used to investigate sensitisation effects. These are the adjuvant method, where a response in at least 30% of the test animals is required in order to claim a positive result and the non-adjuvant method, where at least 15% of the test animals need to show a positive response.

Table 3.4 The three categories and corresponding risk phrases for carcinogenicity, mutagenicity and substances toxic for reproduction.

Health effect	Category 1	Category 2	Category 3
Carcinogen Can either induce cancer or increase its incidence.	*T; R45 May cause cancer*	*T; R45 May cause cancer*	*Xn; R40 Possible risk of irreversible effects*
	T; R49 May cause cancer by inhalation	*T; R49 May cause cancer by inhalation*	
	For substances known to cause cancer in humans.	Sufficient evidence to presume also carcinogenic in humans (based on clear evidence from appropriate animal studies, or other relevant information)	Cause for concern, based on evidence from appropriate animal studies, or other relevant information, but insufficient evidence to place in Category 2.
Mutagen A substance that may either induce heritable genetic effects or increase their incidence.	*T; R46 May cause heritable genetic damage*	*T; R46 May cause heritable genetic damage*	*Xn; R40 Possible risk of irreversible effects*
	For substances known to be mutagenic in humans.	Sufficient evidence to presume also mutagenic in humans (based on clear evidence from appropriate animal studies, or other relevant information)	Cause for concern, in humans owing to possible mutagenic effects from appropriate animal studies, but insufficient evidence to place in Category 2. However, indications of ability to induce mutations in somatic cells, also give rise to placing in this category. It is a warning for possible carcinogenic activity.
Toxic to reproduction Substances that may impair either male or female reproductive function or cause non-inheritable harmful effects in the offspring.	*T; R60 May impair fertility.* For substances known to impair fertility in humans.	*T; R60 May impair fertility.* For substances which should be regarded as if they impair fertility in humans.	*Xn; R62 Possible risk of impaired fertility.* For substances that cause concern, for human fertility.
	T; R61 May cause harm to the unborn child. For substances known to cause developmental toxicity in humans.	*T; R61 May cause harm to the unborn child.* For substances which should be regarded as if they cause developmental toxicity in humans (for both, based on clear evidence from appropriate animal studies, or other relevant information)	*Xn; R63 Possible risk of harm to the unborn.* For substances which cause concern for humans due to possible developmental toxic effects. Based on evidence from appropriate animal studies, or other relevant information.

Note: If there is concern over the effects on lactation by substances that have been classified as toxic for reproduction, the risk phrase *R64 May cause harm to breast fed babies* may be assigned. This is based on criteria as described in the Dangerous Substances Directive.

Source: CHIP (1997) and EEC Directive 67/548/EEC and its amendments, Annex VI.

Table 3.5 Applicable risk phrases for the classification 'Irritant'.

R38 Irritating to skin	A substance is considered to be irritating to the skin if, after an exposure period of 4 hours, significant inflammation of the skin persists for at least 24 hours. The criterion for 'Significant irritation' is laid out in Annex V of the Dangerous Substances Directive.
R36 Irritating to eyes	A substance is considered to be irritating to the eyes if significant ocular lesions occur within 72 hours after exposure and persist for at least 24 hours. The criterion for 'Significant ocular lesions' is laid out in Annex V.
R41 Risk of serious damage to eyes	This risk phrase is used if severe ocular lesions occur within 72 hours after exposure and are present 24 hours or more after instillation of the test material.
R37 Irritating to respiratory system	This is normally based on reversible irritation effects occurring in the upper respiratory tract during general animal toxicological tests. Otherwise, it can be based on human experience.

Source: CHIP (1997) and EEC Directive 67/548/EEC and its amendments, Annex VI.

Table 3.6 Applicable risk phrases for the classification 'Corrosive'.

R35 Causes severe burns	If when applied to healthy, intact animal skin causes full thickness destruction of skin tissue as a result of up to 3 minutes exposure, or the result can be predicted.
R34 Causes burns	If when applied to healthy, intact animal skin causes full thickness destruction of skin tissue as a result of up to 4 hours exposure, or the result can be predicted. This risk phrase is also assigned to organic hydroperoxides, unless evidence proves otherwise.

Source: CHIP (1997) and EEC Directive 67/548/EEC and its amendments, Annex VI.

3.3.4.4.2 *Sensitising by inhalation* Substances or preparations are classified as sensitising, and assigned the symbol Xn, with the risk phrase *R42: May cause sensitisation by inhalation*, if positive results from appropriate animal tests are available. Otherwise evidence based on human experience can be used if this indicates that the chemical substance or preparation has the ability to cause specific respiratory hypersensitivity. This would usually manifest itself in the form of asthma (see Section 1.4). Other criteria have been established for respiratory hypersensitivity and it is well worth reading the DSD for additional information.

3.3.4.5 *Dangerous for the environment* Using the classification 'Dangerous for the Environment' is a means by which awareness can be given to the user of the potential environmental hazards particular substances pose to ecosystems. At the present time the only classification criteria available are those for aquatic environments. Therefore, classification for non-aquatic environments remains based on criteria specific to the aquatic environment.

For the purpose of this book, the focus will be given to criteria used for the aquatic environment. The classification criteria are established on acute effects observed in fish (LC_{50}, 96 hours), daphnia (EC_{50}, 48 hours) and algae (IC_{50}, 72 hours). The criteria are also related to persistence or bioaccumulation potentials (log P_{ow} or B_{CF})

and these values are also taken into consideration for some of the classifications. Table 3.7 gives the classification criteria for the 'Very Toxic', 'Toxic' and 'Harmful' categories, along with the associated risk phrases and symbols.

Table 3.7 Classification and corresponding criteria.

Classification of substance	Very Toxic		Toxic	Harmful	'Catch all' clause
Symbol assigned	N		N	–	–
Risk phrase assigned	R50 + R53	R50	R51 + R53	R52 + R53	R52 and/or R53
Criteria applicable	(1) LC_{50} or EC_{50} or IC50 ≤ 1 mg l^{-1}	LC_{50} or EC_{50} or IC50 ≤ 1 mg l^{-1}	(1) 1 mg l^{-1} $< LC_{50}$ or EC_{50} or IC_{50} ≤ 10 mg l^{-1}	(1) 10 mg l^{-1} $< LC_{50}$ or EC_{50} or IC_{50} ≤ 100 mg l^{-1}	†
	(2) *and* not readily biodegradable *or* log $P_{ow} \geq 3$ (unless $B_{CF} \leq 100$)		(2) *and* not readily biodegradable *or* log $P_{ow} \geq 3$ (unless $B_{CF} \leq 100$)	(2) *and* not readily biodegradable*	

* Other evidence based on biodegradability and chronic toxicity may waive classification.
† Very low water solubility (< 1 mg l^{-1}), not readily biodegradable. A value log $P_{ow} > 3$ may then be assigned in some cases.

3.4 Preparations

A preparation is defined as a mixture of chemical substances. The Dangerous Preparations Directive 88/379/EEC (which will be superseded by 99/45/EC) covers all preparations containing dangerous substances as defined by the Dangerous Substances Directive 67/548/EEC. However, 88/379/EEC is not applicable to preparations that are covered by specific directives such as those for cosmetics and pesticides, etc.

The same basic criteria used in the Dangerous Substances Directive are also applied to the Preparations Directive. However, there is a difference in that the classification of a preparation, which contains one or more dangerous substances, is based on three criteria.

(1) A calculation method that utilises cut-off values or concentration limits. For example, sodium hydroxide has assigned classifications based on different concentration limits. These are shown in Table 3.8. Thus, a preparation containing 3% sodium hydroxide would be classified as *Corrosive, C; R34 Causes burns*. For chemical substances listed in Annex I of the DSD that do not possess any specific concentration limits, certain 'generic limits' have been implemented. For instance, preparations containing more than 0.1% of a substance classified as a Category 1 or 2 carcinogen would be labelled as per

Table 3.8 The different classifications for NaOH based on different concentrations.

Sodium hydroxide	Classification
Concentration ≥ 5%	C; R35
2% ≤ concentration < 5%	C; R34
0.5% ≤ concentration < 2%	Xi; R36/38

that substance; preparations containing a carcinogen at a concentration greater than 1% would be classified as a Category 3 carcinogen. Similarly, if the preparation contains more than 1% of substance classified as a sensitiser, then the preparation is classified as per that substance.

(2) Testing the preparation using Annex V test guidelines.

(3) Reliable human evidence based on detailed and good-quality data.

Results from relevant animal studies can be applied to override a calculation-based classification for a preparation, however, if the information on exposure to humans indicates that the effects will be different from those suggested by the calculation method or testing. The exception is with specific health effects, such as carcinogenicity, mutagenicity and substances that are toxic for reproduction. In these instances the calculation method with cut-off limits has to be adhered to.

3.4.1 *Criteria for the 'Dangerous for the environment' classification*

Up until 1999, criteria used in classifying the environmental effects of preparations have not been available. However, the new Dangerous Preparations Directive 1999/45/EC provides the necessary criteria for the classification 'Dangerous for the environment'.

For classification of ecotoxicological effects, testing on the actual preparation or a calculation method can be used. For parameters such as the persistence and bioaccumulation of a preparation, however, the calculation method must be adopted because these two parameters are considered to be inherent properties of the individual substances and would not change in a preparation.

The Directive 1999/45/EC, which will come into force nationally from 2002, has the proviso that no new animal studies can be carried out unless justified by the fact that intrinsic hazards cannot be sufficiently established by the calculation method or the availability of data from existing animal studies. This should be good news in terms of reducing the number of unnecessary animal studies!

3.5 **Canada**

The 'Workplace Hazardous Materials Information System' (WHMIS) is the Canadian legislation covering the use of hazardous materials in the workplace. Its goal is to ensure that workers are made fully aware of the hazards of chemicals in the workplace and that appropriate training is provided to enable them to work with them safely. WHMIS covers a number aspects including hazard identification, product classification, labelling, and Material Safety Data Sheets together with worker education and training.

3.5.1 *Hazard identification*

There are six different classes of hazard defined within WHMIS, some of which contain sub-divisions. These classes are listed below and the sub-divisions are shown for the class related to health effects, Class D – Poisonous and infectious material. For more information on the other classes, the reader should consult the Controlled Products Regulations outlined in Section 3.5.2.

(1) Class A – Compressed gas
(2) Class B – Flammable and combustible material
(3) Class C – Oxidising material
(4) Class D – Poisonous and infectious material
    (a) Division 1: materials causing immediate and serious toxic effects
        (i) Subdivision A: very toxic material
        (ii) Subdivision B: toxic material
    (b) Division 2: materials causing other toxic effects
        (i) Subdivision A: very toxic material
        (ii) Subdivision B: toxic material
    (c) Division 3: biohazardous infectious material
(5) Class E – Corrosive material
(6) Class F – Dangerously reactive material

3.5.2 *Controlled Products Regulations*

These regulations, which are governed by the federal Hazardous Products Act, outline the hazard criteria used to determine whether or not a chemical should be allocated to one of the classes mentioned above. Chemical substances falling into one or more of these classes are termed *controlled products* and must have an MSDS and be labelled so as to alert the user of the hazard. (Note the WHMIS label has a distinctive frame.)

As with the EU directives, each of the classes mentioned in Section 3.5.1 has a symbol that enables the hazard to be identified at a glance. Under the WHMIS guidelines, risk phrases (different to those used in Europe) and precautionary statements are used to convey hazard information.

For the purpose of this book, only *Class D – Poisonous and infectious material, Divisions 1 and 2* will be summarised (omitting Division 3). Further information on the other classes can be obtained by referring to the Controlled Products Regulations under the Hazardous Products Act.

Tables 3.9 and 3.10 show the criteria for the two divisions and sub-divisions of Class D – Poisonous and infectious material, and are applicable to pure chemical substances or tested mixtures. In this context, the term 'mixture' implies a

Table 3.9 The criteria for Division 1: Materials causing immediate and serious toxic effects, Sub-Divisions A and B, respectively.

Sub-Division A – Very Toxic Material	Sub-Division B – Toxic Material
LD_{50} (oral, rat) $\leq$ 50 mg kg^{-1} (OECD 401)	50 < LD_{50} (oral, rat) $\leq$ 500 mg kg^{-1} (OECD 401)
LD50 (dermal, rat) $\leq$ 200 mg kg^{-1} (OECD 402)	200 < LD_{50} (dermal, rat) < 1000 mg kg^{-1} (OECD 402)
LD_{50} (inhalation, gas/vapour, rat) $\leq$ 1500 ppm; or LD_{50} (inhalation, dust, mist or fume, rat) $\leq$ 50 mg m^{-3} (OECD 403)	1500 < LD_{50} (inhalation, gas/vapour, rat) $\leq$ 2500 ppm; or 0.5 < LD_{50} (inhalation, dust, mist or fume) $\leq$ 2.5 mg m^{-3} (OECD 403)

An *untested mixture* falls into either of the two Sub-Divisions if it contains $\geq$ 1% of a pure substance or tested mixture that fulfils any of the criteria mentioned above.

Source: Controlled Products Regulations.

Table 3.10 The criteria for Class D – Poisonous and infectious material, Division 2, Sub-Divisions A and B, respectively

Criteria	Sub-Division A	Sub-Division B
	Very Toxic Material	Toxic Material
Chronic Toxic Effects (If chronic or sub-chronic rodent/ non-rodent studies indicate either serious irreversible impairment or life-threatening effects at the dose levels shown)	(Oral route, OECD 408, 409, 452) $\leq$ 10 mg kg^{-1}	10 < (Oral route, OECD 408, 409, 452) $\leq$ 100 mg kg^{-1}
	(Dermal route, OECD 411, 452) $\leq$ 20 mg kg^{-1}	20 < (Dermal route, OECD 411, 452) $\leq$ 200 mg kg^{-1}
	(Inhalation route, OECD 413, 452) $\leq$ 25 ppm (volume of gas or vapour), *or*	25 < (Inhalation route, OECD 413,452) $\leq$ 250 ppm (Volume gas or vapour), *or*
	$\leq$ 10 mg m^{-3} (dust, mist or fume)	10 < (Inhalation route, OECD 413, 452) $\leq$ 100 mg m^{-3} (dust, mist or fume)
Carcinogenic Effects	If listed on the most recent versions of: ACGIH (TLV), sections A1a, A1b, A2 of Appendix A IARC listed Group 1 or 2	No criteria given
Teratogenicity and Embryotoxicity	If shown to cause injury (which includes death, malformation, growth retardation) to either the embryo or foetus in a statistically significant proportion of the test population at a concentration which is not maternally toxic as per OECD 414, 415 or 416	No criteria given

Contd

Table 3.10 *(Contd)*

Criteria	Sub-Division A	Sub-Division B
	Very Toxic Material	Toxic Material
Reproductive Toxicity	If there is human evidence of adverse effects on the reproductive ability or causes sterility, arising from exposure in the workplace. Results from animal studies demonstrating these same effects when carrying out a one-generation or two-generation reproductive study as per OECD 415 or OECD 416, respectively.	No criteria given
Respiratory Tract Sensitisation	If there is adequate human evidence to demonstrate that the chemical causes skin sensitisation to those who have been exposed to it in the workplace	No criteria given
Mutagenicity	There is epidemiological evidence available	If positive results are obtained in an *in vivo* mammalian somatic cell test for either gene mutations or chromosomal aberrations
	Positive results from test that measures mutations transmitted to offspring	
	Positive results from tests using *in vivo* germ cells and gene mutation or chromosomal aberrations in somatic cells	
Skin Irritation	No criteria given	If using OECD 404, the following values obtained: 1. A mean value of ≥ 2 for erythema formation 2. A mean value of ≥ 2 for oedema formation (OECD 404)
Eye Irritation	No criteria given	If using OECD 405, the following values obtained: 1. A mean value of ≥ 2 (corneal damage) 2. A mean value of ≥ 1 (damage to the iris) 3. A mean value of ≥ 2.5 (conjunctival swelling or redness)

Contd

Medical Library Service
College of Physicians & Surgeons of B.C.
1383 W. 8th Ave.
Vancouver, B.C.
V6H 4C4

Table 3.10 *(Contd)*

Criteria	Sub-Division A	Sub-Division B
	Very Toxic Material	Toxic Material
Skin Sensitisation	No criteria given	Using OECD 406 test 'skin sensitisation' the following values are obtained: $\geq$ 30% response in test animals using an adjuvant type test $\geq$ 15% response in test animals using a non-adjuvant type test Or, if there is adequate human evidence to demonstrate that the chemical causes skin sensitisation to those who have been exposed to it in the workplace
Untested Mixtures	$\geq$ 1% of a substance or tested mixture that fulfils the criteria of 'Chronic Toxic Effects' or, $\geq$ 0.1% of a substance or tested mixture that fulfils the criteria of 'Teratogenicity and Embryotoxicity', 'Carcinogenicity', 'Reproductive Toxicity', 'Respiratory Tract Sensitisation' or 'Mutagenicity' as defined in this Sub-Division	If it contains $\geq$ 1% of a substance that fulfils any of the criteria mentioned in Sub-Divisions B

combination of two or more chemical substances that do not react with one another. As with the EU criteria, testing can be carried out on either pure chemical substances or on mixtures (called 'preparations' under EU legislation). The WHMIS requires the use of OECD test guidelines, although under some circumstances different test methods are also acceptable.

3.5.3 *Class D – Poisonous and infectious material*

3.5.3.1 *Division 1: materials causing immediate and serious toxic effects*

3.5.3.1.1 *Acute lethality* Division 1, which contains two sub-divisions, A and B, deals primarily with materials (or chemicals in our case) that can cause lethal effects in short-term tests. A pure chemical substance or tested mixture will be placed in one of these two sub-divisions should the results from an acute lethality study correspond to any of the criteria shown in Table 3.9.

3.5.3.2 *Division 2: materials causing other toxic effects* Chemicals that do not cause immediate, serious effects, but result in effects that could still have serious implications for any individual, are placed in this division.

Table 3.10 shows the criteria used to determine whether or not a pure chemical substance or tested mixture should be placed in this division and which of the two sub-divisions are applicable. Note that corrosivity to living tissues is dealt with in Class E – Corrosive material.

3.6 USA

The purpose of the OSHA Hazard Communication Standard 29CFR 1910.1200 is 'to ensure that the hazards of all chemicals produced or imported are evaluated, and that information concerning their hazards is transmitted to employers and employees. This transmittal of information is to be accomplished by means of comprehensive hazard communication programs, which are to include container labelling and other forms of warning, material safety data sheets and employee training'. In other words:

(1) any scientific evidence relating to the hazardous nature of a chemical must be reviewed by the chemical manufacturer/importer in order to determine whether it is hazardous;

(2) if the chemical is hazardous then an MSDS and container labels must be prepared and issued to users;

(3) a written hazard communication programme has to be developed by the employer who also has to ensure that the employees receive training and information concerning the hazards of the chemicals in the workplace.

The hazards mentioned in the above statement can be either *health hazards* or *physical hazards*. Health hazards include toxicity, carcinogenicity, etc. while physical hazards include flammability and oxidising properties, etc.

For the purpose of this book, attention is given only to health hazards. For more information on the physical hazards, the reader is referred to the Hazard Communication Standard (which is quite reader-friendly!).

3.6.1 *Health hazards, evaluation, criteria and definitions*

Under the HCS 29CFR 1910.1200, a health hazard means 'a chemical for which there is statistically significant evidence based on at least one study conducted in accordance with established scientific principles that acute or chronic health effects may occur in exposed employees'.

Although both acute and chronic health effects as defined by ANSI are those most commonly referred to, the Hazard Communication Standard (HCS) makes the point that the document does not necessarily cover all adverse effects possibly encountered from exposure to chemicals. The standard also states that although it is not realistically possible to quantify every possible health effect arising from chemical exposure, it remains important that workers are kept informed and that necessary protective measures are implemented.

If the hazard evaluation demonstrates that one or more of the health definitions contained in Appendix A of the standard are applicable when the criteria provided in Appendix B of the same standard are met, then the chemical is considered to be

Table 3.11 The definitions given in Appendix A and the criteria listed in Appendix B of the Hazard Communication Standard, 29 CFR 1910.1200.

Appendix A – Definitions

Highly Toxic	A chemical is *Highly Toxic* as per this Standard if: LD_{50} (oral, rat) $\leq$ 50 mg kg^{-1} LD_{50} (dermal, rat) $\leq$ 2000 mg kg^{-1} LC_{50} (inhalation, rat) $\leq$ 2000 ppm (gas/vapour)
Toxic	A chemical is *Toxic* as per this Standard if: 50 < LD_{50} (oral, rat) $\leq$ 500 mg kg^{-1} 200 < LD_{50} (dermal, rat) $\leq$ 1000 mg kg^{-1} 200 < LC_{50} (inhalation, rat) $\leq$ 2000 ppm (gas/vapour)
Carcinogen	A chemical is considered to be a *Carcinogen* as per this Standard if: 1. OSHA regulates it as a carcinogen. (Most recent version) 2. It has been evaluated by IARC and is found to be either a carcinogen or potential carcinogen in humans (most recent version) 3. It is listed on the 'Annual Report on Carcinogens' as a carcinogen or potential carcinogen. This report is published annually by the National Toxicology Program (NTP)
Corrosive	A chemical is considered to be *Corrosive* as per this Standard 'if, when tested on the intact skin of albino rabbits by the method described in the US Department of Transportation in appendix A to 49 CFR part 173, it destroys or changes irreversibly the structure of the tissue at the site of contact following an exposure period of 4 hours'
Irritant	A chemical is a *Skin Irritant* as per this Standard 'if when tested on the skin of albino rabbits it results in an empirical score of $\geq$ 5, when using a patch test technique on abraded and intact skin of six albino rabbits for an exposure period of 4 hours' A chemical is an *Eye Irritant* as per this Standard if more than 4 of the 6 albino rabbits used show ulceration/opacity of the cornea, inflammation of the iris or inflammation of the conjunctiva, using the test method described in 16 CFR 1500.42 or another appropriate method
Sensitiser	The standard defines a *Sensitiser* as 'A chemical that causes a substantial proportion of exposed people or animals to develop an allergic reaction in normal tissue after repeated exposure to the chemical'. The test methods recommended include both the adjuvant and non-adjuvant variety
Target Organ Effects	Include the following examples but this is not a definitive list: Hepatotoxins Agents which damage the lung Nephrotoxins Reproductive toxins Neurotoxins Cutaneous hazards Eye hazards Agents which act on the blood or hematopoietic system

---

Appendix B provides the following criteria for doing hazard determinations:

1. Carcinogenicity, listed as carcinogen or probable carcinogen by NTP, IARC, or OSHA is taken as conclusive evidence. (However, all other relevant scientific data have also to be evaluated in accordance with the Standard.)
2. Human data – from epidemiological studies and any relevant case reports of adverse health effects
3. Animal data – results from relevant test data should be used
4. Adequacy and reporting of data – tests using established scientific principles with statistically significant conclusions are acceptable. *In vitro* studies are alone generally not sufficient in making hazard determinations

hazardous under the provisions of the HCS. The definitions of Appendix A and the criteria listed in Appendix B of the standard are summarised in Table 3.11. Note that the standard emphasises that this is not a definitive list and that other relevant data should also be evaluated.

For the procedure of hazard evaluation, the standard states that 'The chemical manufacturer, importer or employer evaluating chemicals shall treat the following sources as establishing that the chemicals listed in them are hazardous'. The sources are 29 CFR Part 1910, Subpart Z, or Threshold Limit Values for Chemical Substances and Physical Agents in the Work Environment (published by the ACGIH). However, for the chemicals given in these sources, it is still necessary to evaluate any hazards associated with them using the requirements laid out in the HCS. The following three sources are also used in establishing whether or not a chemical is either a potential or proven carcinogen: NTP 'Annual Report on Carcinogens'; IARC monographs; and 29 CFR Part 1910, Subpart Z.

Although there are no specific methods required for hazard determination, suppliers still have to demonstrate that they have sufficiently ascertained the hazards as per the criteria laid out in the standard.

3.6.2 How are mixtures dealt with?

For a tested mixture, the result obtained is used directly to determine whether or not it is hazardous. The hazard of an untested mixture is taken as being the sum of all its individual chemical components. According to the HCS, the mixture is assumed to have the same health hazard as any one of its components present at the 1% or more by weight or volume level. However, for those components having been evaluated as presenting a carcinogenic hazard under the criteria of the HCS standard, a presence at the 0.1% or more level in an untested mixture is enough to classify that mixture as presenting a carcinogenic hazard. (This stipulation applies to both the WHMIS and EU regulations.)

Regardless of the '0.1%' and '1%' rules mentioned above, any component present in a mixture at a quantity below these two 'trigger values' with the potential of being released at concentrations exceeding any stipulated air contamination limit (as laid out in OSHA–PEL or ACGIH TLV regulations), or still presents a health hazard at such concentrations will mean that the mixture in which it is contained should be assumed to have the same health hazard.

3.6.3 Labelling requirements under OSHA

As per the Hazardous Communication Standard, the containers of hazardous chemicals must be labelled with the identity of the material and appropriate hazard warnings together with the name and address of the manufacturer. Such labels must be legible and prominently displayed. *There are no specific requirements for size or colour, or any specified text.* All in-plant containers of hazardous chemicals must always be labelled and the labelling system chosen can be that provided by the supplier. Labels must be in English but secondary languages may be also included. This is covered in HCS 29 CFR 1910.1200.f.

The National Fire Protection Association (NFPA 704) and Hazardous Materials Identification System (HMIS) are both examples of labelling systems used to indicate the hazard rating of a chemical. Both systems use a colour code (four colours: red, blue, white and yellow) to indicate different hazards and a number coding (0–4) that ranks the degree of hazard.

The NFPA system uses a diamond shaped symbol while the HMIS system uses a colour bar code. In the HMIS system, a white bar indicates the level of personal protection needed, and rather than a hazard ranking, a letter is used showing the level of protection recommended. With the NFPA system, additional hazard information, such as water reactivity and oxidiser, is provided (see Fig. 3.2).

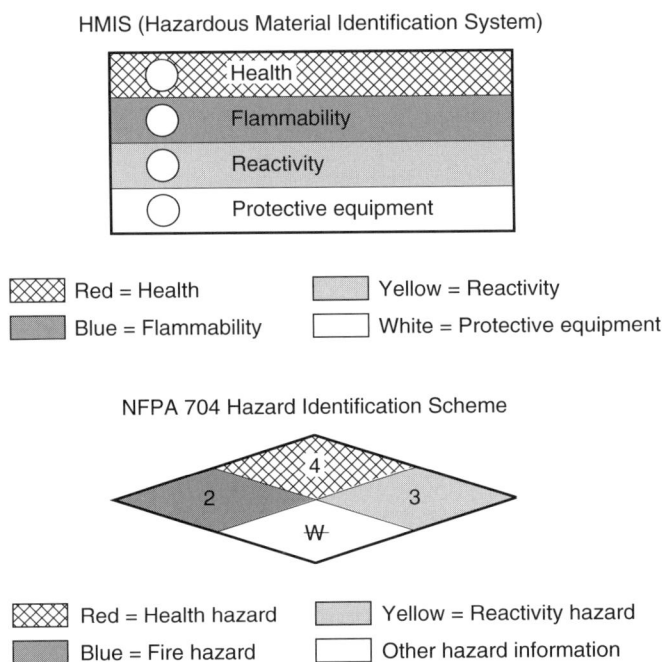

Fig. 3.2 The HMIS and NFPA hazard information systems.

3.7 Summary

The European Union, Canada and the United States have all legislated requirements that the hazardous nature of chemical products is identified and that this information is conveyed to recipient users by means of a label and MSDS.

The criteria used for determining whether or not a chemical product is hazardous can differ to a greater or lesser extent. As an example, consider the criteria for the classification of 'Toxic', as shown in Table 3.12 for the three regulations mentioned in this chapter. Note the differences.

Currently, only Europe has a formal classification scheme for environmental hazards, which is communicated to the user of a chemical via the MSDS and

Table 3.12 The 'Toxic' criteria for the three geographical areas covered in this chapter.

Europe	$25 < LD_{50} \leq 200 \, mg \, kg^{-1}$ (oral, rat)
Canada	$50 < LD_{50} \leq 500 \, mg \, kg^{-1}$ (oral, rat)
USA	$50 < LD_{50} \leq 500 \, mg \, kg^{-1}$ (oral, rat)

container label. However, chemical manufacturers in both the United States and Canada may provide information on environmental effects by way of the so-called 'Responsible Care™ Initiatives'.

Reference

OSHA Hazard Communication Standard 29 CFR 1910.1200

Further reading

OSHA Hazard Communication Standard 29 CFR 1910.1200
Appendix A to 1910.1200 – Health Hazard Definitions (Mandatory)
Appendix B to 1910.1200 – Hazard Definitions (Mandatory)

Classification, Packaging and Labelling of Dangerous Substances in the European Union Part 1
European Commission, January 1997

American National Standard for Hazardous Industrial Chemicals – Precautionary Labelling
ANSI® Z129.1-1994

The Complete Idiot's Guide to CHIP2 or How to Amaze Your Friends with Your Knowledge of CHIP2 without Really Trying
Chemicals (Hazard Information and Packaging for Supply)
HSE Books, 1994

Approved Guide to the Classification and Labelling of Substances and Preparations Dangerous for Supply, 3rd edn
Chemicals (Hazard Information and Packaging for Supply) (Amendment) Regulations
Guidance on Regulations L100
HSE Books, 1997

Useful Internet addresses

Europe

Health and Safety Executive (UK)
http://www.hse.gov.uk

CEFIC Classification and Labelling
http://www.cefic.org/activities/hse/prod_frm.html

VCI Verband der Chemischen Industrie (Germany)
http://www.chemische-industrie.de/

ECETOC (European Centre for Ecotoxicology and Toxicology of Chemicals)
http://db1.nihs.go.jp/ecetoc/

CEFIC–FCA Homepage
http://fca.cefic.org

USA

OSHA Homepage
http://www.OSHA.gov/

ACGIH Homepage
http://www.acgih.org/

NIOSH Homepage
http://www.cdc.gov/niosh/homepage.html

US National Toxicology Program
http://ntp-server.niehs.nih.gov/main_pages/NTP_ARC_PG.html

ICCA HPV tracking system
http://www.icca-che.org/hpv/

Canada
http://www.ccohs.ca/

4 Handling chemicals in the workplace

4.1 Introduction

The previous three chapters presented information about the different aspects of toxicology and ecotoxicology along with the classification criteria used in different countries used to ascertain whether or not a chemical product is hazardous. The aim of this chapter is to assist the recipient of a chemical product in the workplace in utilising this information and to provide some useful 'hot tips' for the user to think about. However, it is not the intention of this chapter to be an all-inclusive replacement for any chemical safety programme.

4.2 Risk assessment

The information concerning the toxicological properties of a chemical that is provided by the supplier or which is obtained from other sources constitutes just one piece of the jigsaw of risk assessment in that it informs the user of that chemical about the intrinsic *hazards* with which it is associated. (See Fig. 4.1 for a summary of this situation.) However, the properties reveal nothing about the *risk* associated with using the chemical. To establish this particular information necessitates a knowledge of the circumstances under which the chemicals are (or will be) used, thus permitting determination of whether or not *exposure* could occur and, if so, how best to eliminate or minimise it. In other words, it is necessary to complete a *risk assessment*.

A risk assessment is simply a way to determine if, how and under what circumstances harm might be caused to an individual. It is a legal requirement in many countries that work situations that involve the use of hazardous chemicals are assessed for associated risks. For the purposes of this book, the terms risk, hazard and exposure are as defined below.

- *Risk* is the likelihood that an adverse effect will occur.
- *Hazard* is an intrinsic property of the chemical that enables it potentially to cause harm. For example, a chemical could be flammable, corrosive, toxic, persistent, etc.
- *Exposure* can be defined as the extent to which a particular hazard is experienced.

As can be seen from Fig. 4.2 there must be both a hazard and also exposure to that hazard in order for there to be a risk of adverse effects. Consider the example of cyanide. As many familiar with murder mystery movies will know, a 'few grains' of cyanide in some victim's drink causes death. However, if the cyanide is stored in a

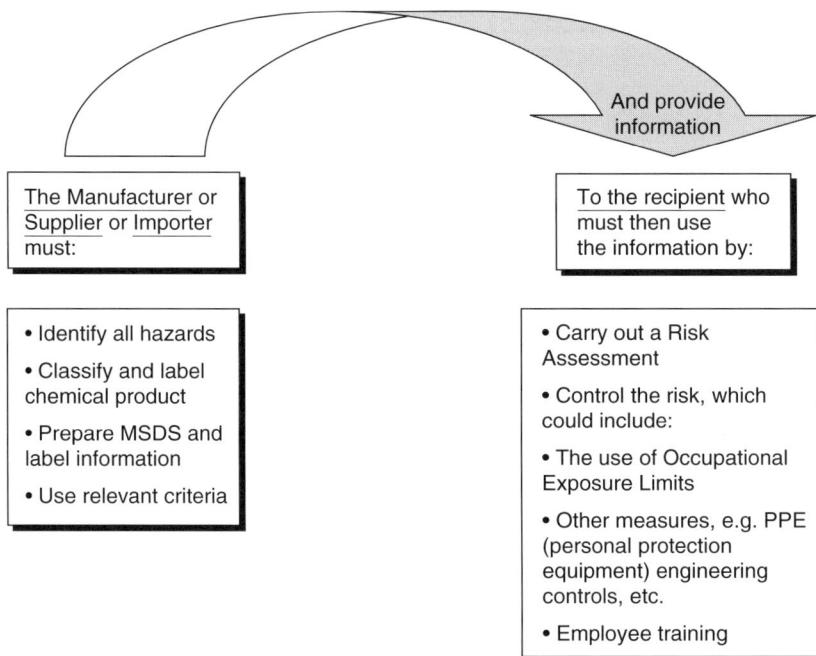

Fig. 4.1 Handling chemicals in the workplace.

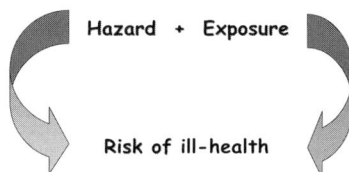

Fig. 4.2 Hazard and risk of adverse effects.

sealed container and is not used, then there is no exposure, which means that there is no risk.

In a nutshell, the purpose of a risk assessment is to answer the question,

'Is this work safe?'

4.2.1 *Some aspects to consider when identifying hazards*

A rather obvious statement to make is that in order to know whether exposure to a hazardous chemical has occurred, or may occur, it is necessary to know what you are looking for. Consequently, an inventory of all chemicals handled on site, regardless of their hazardous nature, is a necessity. Such an inventory is the starting point for all risk assessments associated with the use of chemicals in the workplace. Once the inventory has been prepared, supplementary information regarding the hazardous nature of the chemicals can be added. It is at this point that both toxicological and ecotoxicological information becomes useful, and this can be obtained from the supplier of the chemical via the MSDS or from the supplier's answers to any specific questions. Additional information can also come from literature searches or official publications such as Annex I of the Dangerous Substances Directive in Europe, etc. (A small medical dictionary is invaluable when trying to interpret the medical jargon, which may be used in suppliers' literature!)

When carrying out the process of identifying hazardous chemicals in the workplace, it should not be forgotten that some chemicals can be produced during normal working operations. A good example of this situation involves the fumes that are generated during soldering processes.

4.2.2   Some exposure considerations

In order for an individual to suffer harm, the hazardous chemical has to come into contact with the body in some way, i.e. there has to be *exposure*. There are three main routes by which such exposure can occur.

(1)   *Inhalation*. Lungs are by no means an effective barrier against absorption of chemicals into the body, which is not surprising considering that their prime function is gaseous exchange. Unfortunately, chemicals that are able to reach the lungs and be absorbed into the bloodstream will have a rapid route to other areas of the body. However, because of their particulate shape, other chemicals exert their effect directly in the lungs. An example of this situation is with asbestos dust or crystalline silica Their fibre shape means that both substances can literally 'stick' to the lining of the lungs and not be removed. This causes local chronic inflammatory reactions ultimately resulting in fibrosis which reduces lung function.

(2)   *Skin contact*. The skin is a very effective barrier against most chemicals when it is intact. However, damaged skin will reduce this barrier capacity. Chemicals that are able to penetrate the skin have an ability to cause systemic effects (effects within the body). Of course, local effects such as dermatitis can also occur in some instances.

(3)   *Ingestion*. Providing that good hygiene practices are followed, this should not normally be an exposure route in the workplace, although it can never be totally ruled out. Chemicals that are ingested can either be absorbed from the gut or otherwise pass straight through the body and be excreted.

In general, the physical form of the chemical substance will be of significance when taking into consideration the likely routes of exposure into and the resulting effects on the body.

For example, under certain conditions of exposure a volatile liquid could enter the body both by skin contact and by inhalation of its vapours. In some situations an absorbed chemical can change into a different physical form depending on the environmental conditions, with temperature and air pressure being of particular importance. Therefore, it is important to take into consideration all the possible forms the chemical may assume during its use as well as its form when supplied.

Note also that in some cases a non-hazardous chemical under whichever criteria you choose to adopt could still be an exposure hazard based on its physical form. A good example here would be cationic starch. This chemical is not classified as a health hazard, yet the fact that it is a powder means that should the contamination levels in air be sufficiently high it would, by its physical nature, cause irritation to the eyes and respiratory tract. Therefore, exposure levels need to be controlled and monitored.

The following sections outline some of the physical forms of chemicals that can be encountered when used in the workplace.

4.2.2.1   *Solid*   It can be argued that chemicals that are solid, such as a lump of rosin, are more likely to cause harm by falling on someone's foot than by some other toxic effect! This is not strictly true of course. Although there may not be a risk of poisoning by ingestion, contact with skin could cause either local dermatological problems, such as irritancy, or systemic effects, should molecules be successfully absorbed through the skin.

4.2.2.2   *Dust*   As soon as a solid piece of material is ground into smaller particles a completely different exposure scenario unfolds. The presence of dust in the workplace is both a fire and explosion (dust explosion) hazard. Another problem is that the dust can be inhaled, the outcome on the body depending on the particle size:

(1)   larger particles will not penetrate the lower regions of the respiratory tract, but instead are usually deposited in the nose and trachea/oesophagus where they can be expelled by coughing or sneezing;
(2)   larger particles, together with respiratory secretions, could be ingested ending up in the stomach where they could be absorbed;
(3)   smaller particles or respirable dust can be deposited in the alveoli and respiratory bronchioles of the lungs after inhalation – in some instances this can lead to absorption into the bloodstream and subsequent systemic effects;
(4)   physical irritation can also result from inhaling dusts leading to adverse effects as with asbestos dust, which can cause lung fibrosis.

4.2.2.3   *Liquids*   Liquids categorised as hazardous that are normally stable at room temperature are usually only absorbed through accidental ingestion or contact with the skin in both cases causing either local effects, e.g. stomach upset, dermatitis, or systemic toxicity by absorption into the bloodstream.

4.2.2.4   *Gas*   In terms of acute toxicity effects, it is the gaseous form of a chemical that is potentially its most hazardous. This is because a chemical in the gaseous state can enter the lungs rapidly and be absorbed into the bloodstream usually resulting in immediate systemic effects, or at least local effects in the lungs.

4.2.2.5   *Vapours*   This is a gaseous form of a chemical that is normally found in the solid or liquid state at room temperature and pressure. However, the instability of their normal state under slightly raised temperatures or lowered pressures can readily convert the chemical concerned into a vapour. For example, liquids that have a low boiling point or high vapour pressure can readily volatilise forming vapours that can be inhaled, thereby offering another exposure route to the body. Many organic solvents will evaporate with the production of vapours that can be readily inhaled, or cause irritation to the mucous membranes. Chemicals, solid or liquid, having low boiling points or high vapour pressures are those most likely to rapidly vaporise/volatilise. Thus, it is always useful to read the MSDS or other health and safety literature to ascertain whether vaporisation is likely to occur. If it is, then the inhalation route must also be taken into consideration as a potential route of exposure in the workplace.

4.2.2.6 *Aerosols and mists* An aerosol is a suspension of fine liquid droplets or solid particles in a gas. These droplets or particles are so small that they remain suspended and can become widely distributed. Mists are airborne liquid droplets usually formed by the spraying of liquids, or by the condensation of a chemical from its gaseous to liquid state.

Care must be taken with such chemicals where they are hazardous because skin contact or inhalation will readily occur if individuals are exposed to aerosols or mists containing them.

4.2.2.7 *Fumes* Fumes comprise very small solid particles suspended in air that are formed by the vaporisation of solid chemicals when these are heated to very high temperatures. The problem with fumes is that they are readily inhaled, although owing to their size they are usually exhaled.

4.2.3 *Assessing the exposure*

When attempting to assess the potential level of exposure to a hazardous chemical, a number of factors must be taken into consideration:

- the amount that is normally used in the process and also stored on site
- the frequency with which it is used
- the duration of use along with the number of workers who are potentially exposed.

What should not be forgotten also is any exposure that arises from cleaning (of tanks, etc.) or maintenance work. This is because these two tasks alone can give a much higher level of exposure than the ordinary use of the chemical (taken in the context that the chemical is not specifically used for cleaning purposes).

Although for most hazardous chemicals, such as irritants, the level of exposure will increase the severity of the effects, chemical carcinogens and mutagens are an exception. For these, assessing the level of exposure will influence the probability of contracting a cancer and has nothing to do with the severity of the effects.

4.2.4   *How exposure can be monitored*

Clues other than the advent of adverse health effects, such as the development of a rash, can be used as indicators of exposure. Similarly, it is fairly obvious to state that should a chemical be smelled, then it is being inhaled. Therefore, in some cases, the sense of smell can be a useful indicator that exposure by inhalation has occurred. However, some chemicals do not have any odour, such as carbon dioxide, but can still cause harm, if not death, should the exposure levels be sufficiently high. Also, some chemicals have a low odour threshold. This means that the chemical is readily detectable by smell, though at concentrations below that which could cause harm. Styrene is an example of this. Therefore, the use of odour as an indicator of exposure to a chemical has to be taken into context with other factors, and should not be used alone, especially when the sense of smell can vary widely between individuals. Additionally, ingestion of a chemical can be detected if it has a distinctive taste, a reasonable sign that exposure has occurred.

Should settled dust be present around the workplace and on clothing or hair etc., then it is highly probable that some has also been inhaled whilst it was airborne. This should also be taken as an indication that current exposure control measures are unsatisfactory.

Two common methods by which exposure in the workplace can be measured include biological exposure monitoring and air contamination monitoring.

4.2.5   *Biological exposure monitoring*

This is one form of personal exposure monitoring that involves the measurement of toxic substances or their metabolites in the tissues of the body, fluids such as blood and urine and the exhaled air of workers who have been exposed to a potentially hazardous chemical.

One advantage that biological exposure monitoring has over air contamination monitoring is that it can give an assessment of the uptake by all possible exposure routes, i.e. skin uptake, inhalation and ingestion.

4.2.6   *Air contamination monitoring*

Exposure limits are designed to protect the worker from any adverse health effects arising from *airborne contaminants*. The idea is that provided exposure is kept below the specified limit for the given chemical substance, then the risk of an adverse health effect developing in the majority of workers who are daily exposed in their occupations will be low.

The American Congress of Governmental Industrial Hygienists (ACGIH) has compiled a list of threshold limit values (TLV) that comprised the first set of 'reference value guidelines' against which it was possible to determine the acceptability of measured exposure levels of chemicals in the workplace. Since then, other countries have followed suit and implemented their own exposure limit values, which are either recommendations/guidelines or legally binding directives. Some countries may have

Table 4.1 Occupational exposure limit regulations (various countries).

USA	OSHA-PEL (Permissible Exposure Limit, these are legally binding)
USA	ACGIH-TLV (Threshold Limit Value, these are recommendations)
USA	OSHA-REL (Recommended Exposure Limit)
Germany	MAK (Maximale Arbeitsplatz Konzentrationen)
UK	EH40 OEL (Occupational Exposure Standard)

more than one set of limit values, as is the situation in the United States (see Table 4.1).

Occupational exposure limit values are derived from concentration measurements of hazardous substances present in air that are averaged over a specified period of time. The values are obtained from different sources including toxicological information and industrial experience.

4.2.6.1 *Why do only some chemicals have exposure limits?* Since the objective of monitoring chemicals in the workplace is to protect the worker against exposure to any airborne contamination, only those chemicals that become airborne will be assigned an occupational exposure limit.

Occupational exposure limit values are only guidelines and may not always completely protect workers against the possibility of adverse health effects from hazardous chemicals. This can arise for different reasons including the quality of the data on which the exposure limit value was based, individual differences, multiple chemical interactions and exposure by other routes, etc.

There is no air contamination limit that is 'cast in stone', i.e. values can be reviewed and modified. This kind of revision is usually carried out as a result of new toxicological findings, industrial practical experience, etc. which indicates that the limit in question is, for example, either too strict or conversely, inadequate.

4.2.6.2 *What kind of exposure limits exist?* Essentially there are two kinds of exposure limit.

(1) *Health-based.* For these limits, sufficient evidence is available to demonstrate that provided exposure is kept below the specified limit, there will be no adverse health effects in the majority of workers who are daily exposed to a hazardous chemical. Of course, there will always be a few workers who are more susceptible, perhaps as a result of a predisposing illness or because of hereditary factors. Examples of health-based exposure limit recommendations include the UK EH40 Occupational Exposure Standard (OES) and the German MAK values.

(2) *Non-health-based (technical limits).* These limits are generally compiled for those substances such as carcinogens or respiratory sensitisers where it is not possible to establish an exposure limit that is deemed to be completely free

from that particular risk. This is because these substances at any exposure level, even the smallest amount possible, are potentially hazardous to workers' health. However, since adherence to such levels does not automatically guarantee health protection, it is a prerequisite that every effort is made to reduce the exposure level by as much as possible. There are also other circumstances where a non-health-based standard is implemented, e.g. where there are insufficient data (or unsatisfactory data) available for use in the setting of a health-based limit. Two examples of such a standard are the UK EH40 Maximum Exposure Limit (MEL) and the German Technical Guidance Concentration (TRK).

4.2.6.2.1 *Time periods of exposure* Most occupational exposure systems make the distinction between different time periods of exposure. These are short-term (15 minutes) and long-term (8 hours) exposure limits. The ACGIH (1996) defined both of these as stated below.

(1) Long-term exposures, TLV-TWA. The time-weighted average concentration for a conventional eight-hour workday and a 40-hour work week, to which nearly all workers may be repeatedly exposed, day after day, without adverse effect.
(2) Short-term exposures, TLV-STEL. The short-term exposure limit is a 15 minute TWA exposure, which should not be exceeded at any time during a workday even if the eight-hour TWA is being adhered to. Exposures above the TWA up to the STEL should be no longer than 15 minutes and should not occur more than four times per day.

Short-term exposure limits are typically set for substances where atmospheric concentrations of the substance could be high enough to produce acute effects, such as eye irritation.

Although for some acute effects such as corrosivity, narcotic effects and irritation, a ceiling limit could be allocated. In different countries the above definitions can deviate to some extent and therefore exposure limits may vary. There are also ceiling limits defined for those substances that are fast-acting, e.g. where the main effects are narcosis or corrosion. Overstepping such values for these chemicals is not allowed under any circumstances.

4.2.6.3 *Take note!* In summary consider the following three points.

(1) Occupational exposure limit values for substances are set in order to deal with many different hazards ranging from carcinogenicity to irritation. However, such values cannot be used as relative indicators of a hazard or toxicity, i.e. a high occupational exposure limit value does not necessarily mean that a chemical is more hazardous than another with a lower value – it depends on the reasons for setting the OEL at a particular value. Similarly, two chemical substances with the same OEL does not mean that they present the same level of hazard.
(2) Occupational exposure limit values are not always set purely on the basis of

results from medical monitoring. Limit values can also account for the economic costs of implementing some form of exposure control. Consequently, many occupational exposure limit values are not as low as they should be.

(3) Exposure limit values are usually applicable only to those exposures occurring by inhalation. However, in some instances, chemical substances can be rapidly absorbed through the skin and in order to highlight this fact, a 'skin notation' would be added to those chemicals of concern. In these instances, it is especially important to minimise exposure by skin contact when working with these chemicals.

4.2.7   *How are exposure limit values used in the workplace and which values should be used?*

Put simply, one should adhere to any national exposure limit values that have been issued and are in force in one's place of work. However, if no such values are available, then it would be prudent to obtain at least one set of exposure limit values from those available from monitoring bodies in, for instance, the United States or Europe and then ensure compliance with the lowest values listed. However, some companies use qualified toxicologists and other occupational health specialists to create their own internal standards.

Remember that lists of occupational exposure limits are updated periodically and it is therefore important to acquire the most recent list available. This is especially important if a product is received from a supplier which is accompanied by a Safety Data Sheet that may not give the most up-to-date occupational exposure limit.

If measurements indicate that the occupational exposure limit for workers is being approached, for instance the eight-hour TWA value, then this should be taken as an indication that controls need to be improved in the workplace. Overall, it is important to keep the air contamination limit values in the workplace as low as possible.

4.2.8   *What next?*

Having carried out a risk assessment it is possible to ascertain what precautionary measures are required in order to either eliminate any risk or, where this is not reasonably practicable, to minimise the risk through the use of adequate control measures. These include:

- elimination or substitution
- engineering controls
- work practice and hygiene
- personal protection equipment.

4.3   Chemical hazards

Thus far, risk assessment, hazard identification, routes of exposure and physical form aspects, not forgetting exposure limit values, have all been described. This section provides some extra information about what should be taken into consideration when dealing with different types of health and environmental hazards previously described in the first three chapters of this book.

It is not the intention to go into great detail here. Instead the aim of this section is to simply provide a few additional comments that may be of use. Once again, it is emphasised that the relevant regulations should always be consulted when dealing with hazardous chemicals and, regardless of the nature of the hazard, *all* chemicals should be treated with respect and the appropriate personal precautions taken. One general comment applicable to the handling of chemicals is:

> *'knowledge is the best defence'*

Always read the label and any accompanying MSDS prior to working with the chemical.

4.3.1   *Chemical irritants*

- Chemicals that are known cutaneous irritants – by virtue of test results (Draize test, etc.), structural analogues or based on human experience – will be labelled and have a relevant MSDS so that the user is alerted to the hazard and the appropriate protective precautions that should be taken. Remember that should the chemical have an extreme pH value, namely a fairly high or low pH number, then it will act as an irritant. In this respect, accidental exposure to a primary irritant will almost certainly be felt rather quickly!

- Some chemicals, however, may not be irritants based on the above test methods. Instead their effects are insidious and are exerted by way of cumulative exposure the outcome being dermatitis where protection used has been inadequate. This is because such chemicals are able to defat or degrease the skin in some way. For these chemical types there is no labelling requirement as such, yet in many instances they can be as troublesome as primary irritants. Therefore, always check the supporting documentation, and refer to other literature, to ascertain whether or not any reference has been made to the chemical having the propensity to degrease/defat the skin. Occasionally, but not always, the pH value can be used as an indicator of this behaviour. For example, colloidal silica sol can cause degreasing effects upon repeated contact with the skin.

- Occlusion is another issue that should be considered. As soon as the skin is occluded in some way the penetration rate of a chemical through the upper dermal layers increases dramatically. Therefore, as was mentioned in Section 1.3.1.3, *always* make sure that the skin is clean and dry before wearing gloves.

- For irritants and corrosive chemicals (especially notorious acids and alkalis) take all appropriate measures to avoid the generation of respirable aerosols, mists or dusts – consider how such chemicals are used. For example, should the process in which they are used involve some kind of heating, then it is possible that hot vapours could be generated. Without appropriate engineering measures (such as extractor fans) the vapours would be very irritating to all exposed areas of the body with which they come into contact, especially the mucous membranes.

- Chemicals that are primary skin irritants (as per the Draize test), will usually cause eye irritancy and respiratory tract irritancy, even though they may not be labelled as such (unless there is evidence to the contrary).

With all of the above observations the extent of injury can, at least in part, be reduced by the use of rapid first-aid measures to remove the contaminant.

4.3.2 *Corrosive chemicals*

Chemicals that are corrosive will give rise to severe effects upon contact. The extent of the severity will depend on the length of time that the individual is in contact with the chemical (which also takes into account how quickly it can be washed away). Well-known examples of corrosive chemicals are strong acids and bases.

- As with chemical irritants, it is important to minimise the generation of dusts, aerosols and mists which could be inhaled (thus, take measures to remove corrosive vapours, fumes, dusts or mists).
- If possible, handle corrosive chemicals in pellet rather than powder form.
- Always use corrosive chemicals in well-ventilated areas.
- Inhalation of corrosive chemicals will be severely irritating to the linings of the nose, throat and lungs.
- Eye contact with a corrosive chemical could cause blindness. Therefore, effective and rapid first-aid measures followed by medical attention can greatly minimise the risk of serious, permanent injury. Always read the MSDS, and/or the supplier's label prior to use of any corrosive chemical.

4.3.3 *Sensitisers*

Becoming allergic to a chemical used in the workplace usually has long-term implications. This is because once a worker is sensitised to the chemical, subsequent exposure to even the smallest levels of that specific allergen can trigger an outbreak of allergic symptoms. At best such symptoms are just an 'inconvenience', but at worst the health effects could be life threatening.

The levels of allergen concerned with respect to sensitised individuals can often be so low that normal precautionary measures are insufficient. As a consequence, the only way to avoid an outbreak of allergic symptoms is by completely avoiding all contact with the relevant chemical. Nevertheless, problems do not always end when exposure to that particular allergen is ceased. This is because other chemicals that are structurally similar to the offending antigen are able to provoke an immune response. This response is known as a *cross-sensitivity reaction* and is common with certain chemicals such as acid anhydrides. Even at home it is possible to experience such reactions. The expression 'fur and feathers' is one such case in that those who are unfortunate enough to be allergic to cats (fur) can also be allergic to feathers, or vice versa.

Therefore, handling any chemical where tests have shown it to be a sensitiser should be done with great care, and appropriate precautionary measures should be taken at the outset. Those workers who begin to develop allergic symptoms should ideally not be permitted to work with the chemical in question.

In summary:

- symptoms of respiratory sensitisation may not be manifested directly;
- chemical sensitisers should ideally be substituted for something which 'satisfactorily does the job' *and* which is less hazardous, otherwise any possibility of exposure should be either eliminated or minimised by the use of appropriate control measures;
- once sensitised, very low concentrations far below any air contamination limit value can trigger an attack of symptoms.

4.3.4   *Chemicals that are carcinogens, mutagens or toxic for reproduction*

Chemicals that are carcinogenic, mutagenic or are toxic for reproduction (in the broadest sense) have for all intents and purposes a similar set of usage rules that are applicable should it not be reasonably practicable to use an alternative chemical or process which prevents exposure from occurring. If one is handling any of these chemical types in the workplace, then the national regulations that apply should be consulted.

The following statements expand on precautions to be taken when using these types of chemicals.

(1)   The level of exposure will influence the probability of a cancer occurring and will not influence the severity of the disease itself. This is because it is assumed that there exists no threshold level for carcinogenicity and therefore, a person either develops cancer or does not. As a consequence it is vitally important to minimise absolutely any exposure to a carcinogen – and likewise for chemical mutagens.

(2)   Chemicals toxic for reproduction, i.e. causing adverse effects on either the reproductive process or the developing foetus, are now receiving more attention. One area in particular that is receiving considerable media coverage has been the adverse effect of chemicals on the reproductive ability of males. Unfortunately the effects of many chemicals on the reproductive capability of both males and females and the developing young are unknown. However, this lack of data is being addressed, in part, by the HPV initiative (refer to Chapter 5).

(3)   Different countries have imposed restrictions on women of child-bearing age working with certain chemicals that are thought to cause adverse effects on the unborn child. Relevant local legislation should be consulted.

When handling any chemical which is carcinogenic, mutagenic or toxic for reproduction, substitution of the chemical is the first line of defence. If this is not possible, maintain exposure to a level that is as low as practicably possible. This can be achieved by using totally enclosed systems, or other effective engineering controls, and also by storing chemicals in small amounts. However, should larger amounts of the chemical be needed then it would be best to localise these and use suitably sufficient warning signs.

Additional measures include keeping to a minimum the number of people who could potentially be exposed to these chemicals, monitoring workplace exposure levels and incorporating as routine regular medical checks/health surveillance and a register of those that have used these chemicals in their work. *Always* consult local regulatory requirements or recommendations.

4.3.5 *Toxic chemicals*

Although few chemicals are toxic by all three exposure routes, it could be (and probably is) the case that only one particular exposure route has been investigated and reported. Therefore, be prudent and *never assume* that the other exposure routes present no hazard, unless otherwise stated in 'black and white'.

Having no knowledge about adverse effects arising from a particular exposure route is *not the same* as having knowledge that the adverse effects do not occur.

If the chemical cannot be substituted, keep exposure at a level that is as low as practicably possible in order to measure by air monitoring (if applicable) and maintain regular health checks.

4.3.6 *Environmentally hazardous chemicals*

In Europe there are criteria for the classification of environmentally hazardous chemicals that is based on acute toxicity values, biodegradability and bioaccumulation potentials (see Chapter 2). The following comments are directed more towards the European situation although they are equally valid for other parts of the world, even if no such formal classification and criteria exist.

- Ideally, where possible and appropriate, try to replace all environmentally hazardous materials. Many local authorities will encourage this.
- Contain all environmentally hazardous chemicals in a bunker so that should spillage occur the chemical cannot leak out into the surrounding environment.
- Some pulp and paper mills in Europe prohibit the use of chemicals that are environmentally hazardous. This trend could spread to other areas of the world.
- The new Preparations Directive, which will be implemented at a national level in Europe in 2002, contains criteria for the classification of preparations as 'dangerous for the environment'. This will mean that many preparations that were not previously labelled will now be covered by the new classification criteria even though their compositions have not changed.

Further reading

1996 Tlv®s and BEI®s: Threshold Limit Values for Chemical Substances and Physical Agents – Biological Exposure Indices
Second printing
ACGIH Worldwide

General COSHH ACOP and Carcinogens ACOP and Biological Agents ACOP
Health and Safety Executive, 1999

General and Applied Toxicology, abridged edn
Edited by Bryan Ballantyne, Timothy Marrs & Paul Turner
Macmillan Press, 1995

Occupational Toxicology
Edited by Neill H. Stacey
Taylor & Francis, 1993

Basic Toxicology: Fundamentals, Target Organs and Risk Assessment, 3rd edn
Frank C. Lu
Taylor & Francis, 1996

NIOSH: Pocket Guide to CHEMICAL HAZARDS
US Department of Health and Human Services
Public Health Service Centers for Disease Control and Prevention
National Institute for Occupational Safety and Health, June 1997

Useful Internet addresses

Europe (United Kingdom)

http://www.hse.gov.uk

USA

http://www.OSHA.gov/

http://www.acgih.org/

http://www.cdc.gov/niosh/homepage.html

Canada

http://www.ccohs.ca/

5 Regulatory affairs

DR COLIN C. ROGERS
Regulatory Affairs, Eka Chemicals

5.1 Introduction

In many of the more advanced nations the regulation of chemicals is conducted through quite strict and stringent laws. These regulations fall into two groups or families: there are those which relate to the manufacture and marketing of chemical products and there are those that govern end-use or application (see Table 5.1).

Table 5.1 Chemical regulations.

Manufacturing and marketing	End-use or application
Import/export	Food contact
Existing substances	Ecolabelling
New substances	Toy regulations
Classification, packaging and labelling	Tobacco
	Cosmetics

The basis of these regulations is to protect both humans and the environment from the effects of exposure to chemicals. Chemicals are described in these regulations as *substances*, and mixtures of substances are described as *preparations*. (For a fuller explanation the reader is referred to the Glossary in Chapter 7.)

Many chemical companies are also concerned for humans and the environment and often go beyond their legal obligations in their activities and subscribe to the principles of Responsible Care™. In very simple terms, this means that producers of chemical products take care of their products during the entire life cycle, and attempt to manage the safe handling and use by all concerned in the use-chain, i.e. from cradle to grave. One of the pillars of Responsible Care™ is Product Stewardship. In practising this, producers demonstrate concern for the health and welfare of the users of their chemical products. Companies practising Product Stewardship are proactive in providing information about the health, safety and environmental impact arising from the use of their chemical products.

5.2 Chemical inventories

Table 5.2 shows a list of the most well-known national inventories included in legislation relating to the manufacture *and importation* of chemical substances. It is important to understand that these laws apply equally to producers as well as importers. If importing a chemical product from overseas then these laws must be

Table 5.2   National inventories for the control of hazardous chemicals.

Country	Inventory
Europe	EINECS
USA	TSCA
Canada	DSL
Australia	AICS
China	SEPA
Korea	KECL
Japan	MITI
Philippines	PICCS

observed. Suppliers will of course help with the provision of essential information to comply with the law. Only those substances listed are allowed legally to be placed on the market in those countries. Such substances are known as 'existing substances', and are present on the market in that country. However, the listing of a substance on an inventory in one country does not automatically mean that it will be listed on the inventory in another country. Each inventory is compiled using a unique set of rules. By way of illustration, in Europe, polymers such as polyacrylamide are exempt from listing on the EINECS inventory because acrylamide monomer is listed on EINECS, whereas elsewhere both have to be specifically listed. The substances are listed using proper chemical names, either taken from the IUPAC nomenclature system or from the Chemical Abstracts Service (CAS) Registry.

Generally, these inventories contain the substances listed in two parts: the public inventory and the confidential listing. As the names imply, the public inventory is published officially whereas the confidential listing is only available to the regulating authority.

In many places producers and importers of chemical substances appearing on the relevant inventory must submit information to the authorities about that substance under the regulations applying to existing substances. The information that must be submitted includes data on volumes made or imported and use patterns, together with data on physical and chemical properties as well as toxicological and ecotoxicological properties. The technical information submitted is published by the authorities.

Should a substance not appear on the national inventory then it becomes a 'New Substance', and if a company wishes to market it in that country it must adhere to the regulations governing the notification of new substances. This entails providing the authorities with a comprehensive package of data on that new substance, similar to the information required under The Notification of New Substances SI 1993 3050. Only after the authorities have reviewed the information and agreed to its manufacture in that country or importation can the company proceed. Penalties for non-compliance can be very severe in financial terms, and can run into millions of dollars.

5.3   Food contact regulations

Many producers of paper products recognise the importance of complying with food contact paper and board regulations, but these are by far the most complex sets of rules, and the least understood. The most commonly cited are those related to Germany, the Netherlands and the USA. Finland also has a regulation specific to paper and board. Other countries have laws governing food contact materials in general as a means of protecting the consumer, and indirectly apply these to paper and board, such as the laws in France, Belgium and Italy. But in Europe there is the principle of 'Mutual Recognition' as provided in the Treaty of Rome (Montford, 1966). This basically states that a product legally fulfilling the requirements in one member state may be legally marketed in another member state providing it does not endanger human health.

It is a popular misconception that substances complying with the various food contact regulations are safe to use for the manufacture of paper and board. Actually, the safety aspects of chemical products are encompassed in the Safety Data Sheets that accompany them. The food contact regulations are actually specific to the paper having been made with the chemicals concerned. They do not, however, relate to the safe use of those chemicals in the paper-mill environment. A summary of the more frequently quoted food contact paper and board regulations is given in Table 5.3.

Table 5.3 Food contact regulations.

Country	Regulation
Germany	BgVV Rec. XXXVI
The Netherlands	VGB Ch. II
USA	21 CFR.176.170
Finland	Decree 539/91

5.3.1 *Germany*

In Germany, the law governing food contact materials is the LMBG (Lebensmittel & Bedarfsgegenstandesgesetz, 1974). Sections 30 and 31 relate to food packaging materials. In essence these state that the material must not contain prohibited substances and must not endanger human health, nor affect the food properties (taste, odour, colour, etc.).

A most common mistake is to regard the recommendations made by the government institute BgVV (Bundesinstitut für gesundheitlichen Verbraucherschutz und Veterinärmedizin) as law. In reality the recommendations only concern the substances regarded as safe for use in the manufacture of food contact materials. In total 52 individual recommendations have been published (Wieczorek & Pfaff, 1999) with Recommendation XXXVI relating to paper and board for general-purpose food contact. For more specific applications, there are Recommendations XXXVI/1 for hot and cold filtration (e.g. tea bags and coffee filters) and XXXVI/2 for bakery papers. Individual substances are listed on this 'positive' list together with certain limitations or restrictions. There are a considerable number of substances not listed such as process aids and formulation aids. These substances are not intended to become part of the finished article and should not be detectable. Such substances, for example, are solvents, and dispersing agents.

The recommendations are constantly being updated as the Plastics Commission of the BgVV adds new substances following evaluation. Often it is possible to obtain an independent expert opinion to certify that a particular chemical product is safe for use in the production of food contact paper that complies with the LMBG and a specific recommendation of the BgVV. In former times the BgVV was known as the BGA (Bundesgesundheitsamt).

5.3.2 The Netherlands

The Netherlands developed independently its own food contact legislation, Verpakkingen en gebruiksartikelenbesluit (VGB, 1999), under the Consumer Protection Act (Warenwet). In Europe it is the most comprehensive and complete set of rules relating to food contact paper and board. Whereas the LMBG of Germany only states the principal substances, the Dutch law actually includes a true positive list of permitted substances, together with the appropriate restrictions and limitations.

In the VGB, there are ten chapters each relating to certain food contact materials. Chapter II is specific to paper and board; with Part 1 for general-purpose food contact circumstances and Part 2 specific to applications above 80°C. These chapters are constantly being updated as the G4 Committee of the Health Ministry adds new substances following evaluation.

5.3.3 United States

Food contact materials in the USA are subject to the Food Drug and Cosmetic Act, and are regulated by the Food & Drug Administration (FDA). A term often used is 'FDA Approved'. The FDA actually strongly objects to such claims. All chemicals that become part of food, whether by accident through migration from the packaging, or design such as a direct food additive, are considered to be 'food additives'. They are either direct or indirect food additives. Each chemical substance is regulated by the FDA and these regulations are consolidated in the well-known Code of Federal Regulations (CFR, 1999). Title 21 of the Code, Parts 170–199, relates specifically to food additives.

Within this Title 21, paragraph 176.170 regulates paper and board for contact with moist and fatty foods, and 176.180 regulates paper and board for dry food contact. It is essential to be conversant with the actual text at the beginning of these paragraphs, because there are cross-references to other paragraphs. Direct food additives are also permitted as ingredients of food contact paper and board, and certain groups of substances such as biocides, defoamers and rosin derivatives are regulated under other paragraphs. Thus, paragraphs 176.170 and 176.180 do not list all the permitted additives for paper and board.

There is also another unique feature of the US Food Drug and Cosmetic Act with which the reader should be aware. If a substance becomes part of the *diet* at a level of less than 0.5 ppb and is not a proven or suspected carcinogen, then it is not subject to regulation by the FDA. This is very important because in a number of instances, such as dry food contact, the actual use does not allow the substance to migrate at levels exceeding this 'Threshold of Regulation'. It is not essential to obtain the sanction of the FDA in such a situation and self-certification is allowed. However, it is possible to obtain an independent opinion from an expert as to whether a substance would comply with the US Food Drug and Cosmetic Act.

5.4 Ecolabelling

Ecolabelling is a good example of 'grey legislation'. Many schemes exist and they all have the common feature in that they relate to consumer products. Paper and board products are included in the many categories, which range from washing machines to household products. Each country operates its own scheme and they all have in common the assessment of the environmental impact of their manufacture and use in practice. Chemicals themselves do not qualify for such an ecolabel, and in some cases are specifically excluded. However, the uses of certain chemicals for the manufacture of paper products may assist the paper producer to qualify for such an ecolabel award. The assessment criteria vary enormously from scheme to scheme, and generalisations are not possible in the space available here. As an illustration, however, the German Blue Angel excludes substances that are classified as carcinogenic for example, and the EU scheme does not make any reference to the chemicals used. This scheme is more related to the production circumstances and the assessments include those of chlorine-containing discharges and energy efficiency. The chemicals used will obviously have an impact on these parameters. A summary of the ecolabelling schemes commonly featured in the paper-making industry is given in Table 5.4.

Table 5.4 Ecolabelling schemes.

Country	Scheme
Nordic countries	White Swan
Germany	Blue Angel
EU	Ecolabel
Sweden	Svensk naturskyddsföreningen

5.5 HPV programmes

Resulting from public pressure, which recognises that many substances in general use do not have comprehensive data on their toxicological properties, various initiatives have been created under the heading HPV. Chemicals produced in High Production Volumes (HPV) have been selected for high-priority action by industry to generate the so-called missing data.

There are two independent initiatives, one operated by the US Environmental Protection Agency (EPA) and the other operated by the International Council of Chemical Associations (ICCA). The former has produced a list of 2800 substances, while the ICCA has a priority list of 1000 substances. There are a number of features common to the two initiatives. Both use the common set of data points covering the physico-chemical toxicological and ecotoxicological properties. This set of data is called the Standard Information Data Set (SIDS), and when the data are complete they are published on the Internet. The data will be presented in summary form and are purely hazard-based. There will not be in the short term any form of risk assessment, although robust summaries of the hazard will be included.

Industry has been galvanised into action and the testing programmes have already begun, with the aim to complete the SIDS for each substance by 2004.

5.6   Issues of concern

There is a number of issues of concern to paper makers, and other users of chemicals in industry. These have been called 'hot issues' because they flare up and then subside as more knowledge becomes available and people realise that their initial fears have become allayed.

5.6.1   Bovine spongiform encephalopathy (BSE)

Many chemicals used in the manufacture of paper are produced from raw materials of natural origin. Such raw materials are fatty acids, and an important source is tallow. Tallow originates from animal fat and there were concerns over the risk of transmission of bovine spongiform encepholopathy (BSE) and similar infective conditions as a result.

The facts are that the agent responsible for the transmission of these diseases, prions, stay with the proteinaceous parts of the body, and in particular the nervous tissues. Experiments were conducted with mice fed tallow from infected cattle. The mice did not become infected.

It is also argued that the various processes used in the production of fatty acids from tallow are sufficient to destroy any prions. High temperatures are used to hydrolyse tallow to the fatty acids and the hydrogenation step would certainly degrade any residual prions.

5.6.2   Genetically modified organisms (GMOs)

There is a considerable concern over the effects on human health and the environment by the introduction of genetically modified organisms into our diet. Many crops are being genetically modified (GM) for various reasons: tomatoes, soya, rape seed, maize, potatoes, etc. There are perceived fears about the long-term health effects. In the paper industry large quantities of starch are used and the sources may well have been genetically modified. A large proportion of the maize grown in the United States is GM, and there is no attempt to segregate it from non-GM crops. In Europe the starch industry has introduced a code of practice: a very small amount of GM maize has been grown, but has not been used for the production of industrial starch. There is now a traceability scheme whereby shipments of crops can be traced back to the exact fields where they were grown and harvested.

Table 5.5 Japanese list of known or potential endocrine disrupters.

Rank	Chemical	Rank	Chemical
1	Dioxin	35	Trifluralin
2	Polychlorinated Biphenyl (PCB)	36	Alkylphenol, Nonylphenol, Octylphenol-4
3	Polybrominated Biphenyl (PBB)	37	Bisphenol-A
4	Hexachlorobenzene (HCB)	38	Diethylhexyl Phthalate
5	Pentachlorophenol (PCP)	39	Butylbenzyl Phthalate
6	Trichlorophenoxy Acetic Acid (2,4,5-T)	40	Dibutyl Phthalate
7	Dichlorophenoxy Acetic Acid (2,4-D)	41	Dicyclohexyl Phthalate
8	Amitrole	42	Diethyl Phthalate
9	Atrazine	43	Benzopyrene
10	Alachlor	44	2,4-Dichlorophenol
11	Simazine	45	Adipic Acid Diethylhexyl Diester
12	Hexachlorocyclohexane	46	Benzophenone
13	Carbyl	47	4-Nitrotoluene
14	Chlordane	48	Octachlorostyrene
15	Oxychlordane	49	Aldicarb
16	Trans-Nonachlor	50	Benomyl
17	1,2-Dibromo-3-chloropropane	51	Kepone
18	DDT	52	Manzeb
19	DDE & DDD	53	Maneb
20	Kelthane	54	Metiram
21	Aldrin	55	Metribuzin
22	Endrin	56	Cypermethrin
23	Dieldrin	57	Esfenvalerate
24	Endosulfan	58	Fenvalerate
25	Heptachlor	59	Permethrin
26	Heptachlor Epoxide	60	Vinclozololin
27	Malathion	61	Zineb
28	Mesomyl	62	Ziram
29	Methoxychlor	63	Dipentyl Phthalate
30	Mirex	64	Dihexyl Phthalate
31	Nitrophene	65	Dipropyl Phthalate
32	Toxaphene	66	Styrene Dimers and Styrene Trimers
33	Tributyltin	67	n-Butylbenzene
34	Triphenyltin		

European maize is for all intents and purposes GM free. Experiments are ongoing with potatoes, so GM potato starch may be produced in the future.

5.6.3 *Endocrine disrupters (EDs)*

Endocrine disrupters (EDs) have a variety of other names. The chemicals concerned, however, all have the same ability of affecting the male reproductive system and bringing about the feminisation of males. This phenomenon has been recognised in a number of wildlife species: fish, alligators, birds, etc. It is not known precisely how these substances affect the male, but they appear to disrupt the endocrine system. There is no definitive list of endocrine disrupting chemicals because there is no validated method of testing for such properties. Many agencies have drawn up lists of suspected or known endocrine disrupters, but one of the most widely circulated and used is the so-called '67 list' (Table 5.5) from the Japanese Environment Agency (1998).

The European Chemical Industry (CEFIC) is funding research into the subject to devise suitable test methodology and is working with the European Commission to draw up an official list, but this is not yet available.

Such chemicals are ubiquitous, and are contained in materials all around us in the Western world: as plasticisers in plastics; as industrial detergents; as surfactants in paints; as components in cosmetics. These substances all have at least one feature in common: they are usually very persistent in the environment and are not easily degraded biotically. Owing to the pressures from environmentalists, producers of paper chemicals no longer use such compounds in their products.

References

CFR (1999) *Code of Federal Regulations*, Title 21, Parts 1–199. US Government Printing Office, Washington, DC.

Japanese Environment Agency (1998) *Strategic Programme on Environmental Disrupters 1998* (SPEED 98). JEA, Tokyo.

Lebensmittel & Bedarfsgegenstandesgesetz (1974) *Bundesgesetzblatt* No. 95 20.8.1974, p. 1945 (as amended).

Montford, J.-P. (1966) The Article 30 solution: an alternative to market food contact materials in the European Union. *Food and Drug Law Journal*, 51(1).

Verpakkingen en Gebruiksartikelenbesluit (VGB) (1999) *Warenwet*, 12th edn. Staatsuitgeverij, DenHaag.

Wieczorek, H. & Pfaff, K., eds (1999) *Kunststoffe im Lebensmittelverkehr*, 49th edn. Carl Heymanns Verlag KG, Cologne.

6 The Material Safety Data Sheet

6.1 Introduction

For any hazardous chemical product being used in the workplace, the Material Safety Data Sheet (MSDS) is the most important document to have accessible. This is because it provides important information related to:

- the safe handling and use of the chemical product (i.e. how to prevent hazardous situations from occurring);
- any potential adverse health and environmental effects arising from exposure to the chemical; and
- the appropriate measures that should be taken in the event of exposure.

Where the English version of this document is used it is referred to as either the MSDS (Material Safety Data Sheet) or the SDS (Safety Data Sheet). In this chapter, the term MSDS will be used to mean both.

The MSDS is essentially a summary of information that has been obtained from different sources about the particular chemical that it covers. Its contents are not an assessment of the risk associated with using the chemical. This is for the simple reason that the supplier of a chemical product covered by an MSDS could never anticipate all possible exposure scenarios arising from its use. Instead, the MSDS provides information related to intrinsic hazards associated with the chemical. These can then be scrutinised for assessments of risk in the workplace.

One very important aspect to note is that an MSDS should *always* be accessible to those using the chemical product to which it pertains.

6.1.1 *When should an MSDS be supplied?*

Many countries have similar legal requirements for instances when an MSDS should be issued. For Europe, Canada and the USA, an MSDS must be supplied if the chemical (substance or preparation/mixture) is classified as hazardous. (Note that in this context, hazardous and dangerous are synonymous; see Chapter 3.) Therefore, if a chemical is not deemed to be hazardous then an MSDS does not have to be provided.

However, many national workplace laws require the provision of documentation that outlines health and safety precautions that should be taken when handling chemicals in the workplace. Consequently, many companies provide an MSDS for all chemicals intended to be used in an occupational setting (occupational setting in the context of this book is the paper-making plant or research laboratory, etc.), regardless of their hazardous nature.

6.1.2 *How is the information organised?*

The information contained within the MSDS is organised into different categories or headings. Most MSDS tend to follow the ISO 11014-1, the EU Directive 93/112/EC or the ANSI Z400.1-1993 standard 16-heading format and all have essentially the same layout. However, it is not uncommon to see an MSDS which has only nine headings. This occurs with the OSHA (USA) and WHMIS (Canada) regulations, although both of these countries will accept the 16-heading format provided certain conditions are fulfilled. Note that the classification criteria used in the MSDS should comply with the national legislation of the country where the product is to be used. Therefore, for a hazardous chemical which is being shipped from the United States to Europe, the company must provide an MSDS using the criteria as laid down in the Dangerous Substances or Dangerous Preparations Directives of the European Union. Since international trade is becoming more prevalent, it is not uncommon for companies to provide MSDS which fulfil the criteria of classification for all countries to which the chemical is being shipped. For example, a US supplier selling a product which is used both in the United States and Europe may well provide an MSDS that satisfies the necessary criteria as laid out in the respective recommendations for both. An example of the 16-heading format MSDS is shown in Fig. 6.1.

As can be seen from this example, the categories are as follows:

(1) Chemical product and company identification
(2) Composition/information on ingredients
(3) Hazards identification
(4) First-aid measures
(5) Fire-fighting measures
(6) Accidental release measures
(7) Handling and storage
(8) Exposure controls/personal protection
(9) Physical and chemical properties
(10) Stability and reactivity
(11) Toxicological information
(12) Ecological information
(13) Disposal considerations
(14) Transport information
(15) Regulatory information
(16) Other information

In the following sections a short summary of the information that should be found under each of the 16 headings is given. Each has been written without any specific country in mind and as such the text should not be used in place of any country-specific legislation regarding the production, content and use of the MSDS.

Since the purpose of the MSDS is the same regardless of the geographical part of the world from which it may have originated, this summary is applicable for wherever the MSDS is being used with respect to the chemical it covers.

Upon receipt of an MSDS for the first time special attention should be paid to sections 9, 10, 11 and 12 as these form the basis for the rest of the MSDS.

AKZO NOBEL

SDS
SAFETY DATA SHEET

1. PRODUCT AND COMPANY IDENTIFICATION

Chemical name: **SODIUM CHLORATE, NaClO₃**

Producer:
Eka Chemicals AB
Bleaching Chemicals Division
SE-445 80 Bohus
SWEDEN

Tel: +46 31 58 70 00
Fax: +46 31 58 78 57

See section 16, about production units

2. COMPOSITION / INFORMATION ON INGREDIENTS

Chemical name	Concentration	CAS number	EC number	HAZCHEM	Hazard labelling	R-phrases [1]
Sodium Chlorate	≥99.5%	7775-09-9	231-887-4	1SE	O (Oxidizing)	R9
					Xn (harmful)	R22

[1] see section 15

3. HAZARDS IDENTIFICATION

Human health effects: Harmful if swallowed.

Environmental effects: Harmful to plants in general, however toxic to brown algae.

Specific hazards:
Oxidizing.
Explosive when mixed with combustible materials.
Strong acids react violently with chlorates, producing toxic and explosive gases, e.g. chlorine and chlorine dioxide.

4. FIRST-AID MEASURES

Inhalation:
Remove to fresh air.
Seek medical care immediately if toxic symptoms occur.

Skin contact:
Wash skin with soap and water.
Remove and rinse contaminated clothing immediately with water.

Eye contact:
Irrigate immediately with water.
Seek medical care if symptoms remain.

Ingestion:
Rinse mouth and drink immediately two or more glasses of water or milk. Do **not** induce vomiting!
Seek medical care if more than a negligible amount has been swallowed.
Risk of methaemoglobinemia. Not to be treated with methylthionin.

Contact ERC, Emergency Response Center for more detailed information tel. +46 8 33 70 43.

5. FIRE-FIGHTING MEASURES

Extinguishing media: Use water as extinguishing medium.

Extinguishing media that are not suitable: Do **not** use powder or CO₂-type extinguishers!

Specific hazards:
A fire can cause explosion when sodium chlorate is present. A confined container with sodium chlorate can explode if heated above the decomposition temperature (>250°C).

Specific methods:
Extinguish fire with large amounts of **water**. Do **not** smother the fire with a blanket! Containers near heat sources must be removed immediately or cooled with water.

Fig. 6.1 An example: the Safety Data Sheet for sodium chlorate.

6. ACCIDENTAL RELEASE MEASURES

Personal precautions: Personal protective equipment (see section 8) is necessary when taking care of a large spillage. Sources of ignition must be kept away.

Environmental precautions: Waste on the ground must be collected.
Solutions should not be drained in sewage systems.

Methods of cleaning up: Dam with sand or similar and collect in a plastic or metal container for disposal.
Do not use sawdust, peat, bark or other combustible materials.
Flush with large amounts of water. Contact experts in case of large spillage.
Notify appropriate authorities.

7. HANDLING AND STORAGE

Handling: Technical measures
Use only inert lubricants and packings for pumps, valves and other equipment when handling chlorate or chlorate solution. Exchange lubricants at regular intervals.

Safe handling advice
- Electrical supply and distribution points are to be cleaned periodically from dust.
- Avoid contact with the incompatible materials given below (incompatible products).
- Keep containers closed when not in use.
- Avoid contact with skin or eyes.
- Keep away from sources of heat and ignition.

Precautions for safe handling
- Chlorates should be handled so as to avoid the scattering of dust.
- Any operation that may cause dust should be ventilated with a suction fan.
- The dust should be collected and disposed of carefully.

Storage: Technical measures
See below.

Storage conditions
- Store in a cool, dry and fireproof area away from sources of heat and ignition, including friction and impact.
- Keep away from food and beverages.

Big bags:
- Store on gravel or crushed stone. Avoid storing in an asphalt paved area.
- The separation between stacks should be at least 8 - 10 m.
- Maintain a safe distance from buildings.

Bulk:
- Store in permanent bins of non-combustible construction.
- Store separate from all other materials.

Incompatible products
Sodium chlorate should be stored separate from organic substances, strong acids, phosphorus, sulphur, sulphides, powdered metals, ammonium salts.

Safe packaging materials
Use UN-approved metal drums or plastic lined bags. Do not reuse big bags.

8. EXPOSURE CONTROLS / PERSONAL PROTECTION

Technical measures:
- Local mechanical ventilation.

Personal protective equipment:
- Respiratory dust filter in dusty atmosphere.
- Gloves, boots, apron of plastic or rubber or protective clothing.
- Goggles giving complete protection to eyes.
- Eye wash bottle with clean water.
- Working clothes should be washed daily in water-based laundry systems.
- Immediately change clothes in any case of contamination of chlorate!

Precaution measures:	Emergency showers or bath tubs filled with water must be available. Impregnated material must be taken care of so that there is no risk of uncontrolled fire.
	NO SMOKING!

9. PHYSICAL AND CHEMICAL PROPERTIES

Physical state and odour:	White crystals, odourless, bitter saline taste
pH in solution:	neutral
Melting point:	248 - 250°C
Boiling point:	decomposes
Decomposition:	250 - 300°C
Flash point:	not applicable
Explosion properties:	see section 10
Density, bulk:	1500 kg/m^3
Solubility in water:	728 g/l at 20°C

10. STABILITY AND REACTIVITY

Sodium chlorate is a strong oxidizing agent and decomposes when heated above 250°C. During the decomposition sodium chloride and oxygen are formed.

Conditions to avoid:	• High temperature
Materials to avoid/ Hazardous decomposition products	• Mixtures of NaClO$_3$ and organic material are very flammable.
	• Explosions can occur when mixed with combustible materials and subjected to addition of energy, e.g. friction or impact.
	• Textiles, cellulose or leather contaminated with chlorate ignite easily.
	• Strong acids react violently with chlorates producing toxic and explosive gases, e.g. chlorine and chlorine dioxide.

11. TOXICOLOGICAL INFORMATION

Sodium chlorate is classified as a harmful substance.

Acute toxicity:	LD$_{50}$ (oral, rat) = 1200 mg/kg
	LD$_{LO}$ (oral, human) = 214 mg/kg
	Sodium chlorate has an oxidizing effect. It oxidizes haemoglobin in the blood to methaemoglobin which has a reduced ability to transport oxygen. This leads to lack of oxygen in the body tissue.
	Symptoms
	Gastroenteric pain, nausea, vomiting, diarrhoea, dyspnoea, cyanosis, acute nephritis, anuria, kidney injury, liver injuries, convulsions, coma, death. The first symptoms may occur after several hours.
Local effects:	Inhalation
	Inhalation of dust causes irritation of the mucous membranes.
	Skin contact
	Sodium chlorate is a mild irritant.
	Eye contact
	Sodium chlorate is a mild irritant.

12. ECOLOGICAL INFORMATION

Chlorate is harmful to plants in general, however toxic to brown algae. Bacteria involved in ammonification, nitrification and denitrification are the most sensitive towards sodium chlorate.

Mobility:	Water
	Stays dissolved in water.
	Soil
	Can be leached out from soil.
Persistence/degradability:	Slow degradation in soil in aerobic conditions. More rapid degradation to sodium chloride and oxygen in anaerobic conditions (microbial degradation).

Bioaccumulation:	Chlorate is converted into chlorite in plants. Chlorite is accumulated in the cells until toxic concentration is reached and the plant dies. There is no evidence of accumulation in animals.
Ecotoxicity:	Harmful to aquatic organisms especially to algae, such as *Fucus vesiculosus* (bladder wrack) for which sodium chlorate is toxic.
	Chlorate can disturb micro-organisms in sediment, for example those involved in the nitrogen cycle.
Fish:	48h LD_{50} *Oncorhynchus mykiss* (rainbow trout) = 2750 mg/l
Crustacean:	24h LC_{50} *Daphnia magna* (water flea) = 880 mg (potassium chlorate)
Algae:	IC_{50}, *Fucus vesiculosus* (bladder wrack) = 0,080 mg/l (long term study). Clear inhibitory effect for *Scenedesmus* (planktonic green algae) at 7mg/l.

13. DISPOSAL CONSIDERATIONS

Waste:	Sodium chlorate is not to be thrown on landfill or in water courses. Collect in plastic or metal containers.
Contaminated packaging:	Remove residues before disposal. Combustible packaging materials, e.g. plastic bags can be burnt in an open vessel under controlled conditions.
General:	Consult appropriate local authorities to ascertain proper disposal procedure.
	Contact Eka Chemicals if technical assistance is needed, see section 16.

14. TRANSPORT INFORMATION

<u>International regulations</u>

UN-no:	1495
Proper shipping name:	SODIUM CHLORATE
Packaging group:	II
Label:	5.1
Land:	ADR/RID Class 5.1 item 11 (b)
Sea:	IMDG Class 5.1 EmS no. 5.1-06 MFAG 745
Air:	IATA-DGR Class 5.1

15. REGULATORY INFORMATION

Classification and labelling:
According to the commission directive 67/548/EEC, 18th technical adaptation, classification and labelling of dangerous substances.

Class: O (oxidizing)
           Xn (harmful)

Labelling:

Oxidizing   **SODIUM CHLORATE**   Harmful

R-phrases (R9-22-31)
Explosive when mixed with combustible material.
Harmful if swallowed.
Contact with acids liberates toxic gases).[1]

S-phrases (S(2)-13-17-46)
Keep out of reach of children).[2]
Keep away from food, drink and animal feeding stuffs.
Keep away from combustible material.
If swallowed, seek medical advice immediately and show this container or label.

[1] Labelling according to Eka Chemicals
[2] S2-label only for consumer goods

16. OTHER INFORMATION

Recommended use
The dominating use of sodium chlorate is generation of chlorine dioxide for bleaching of chemical pulp. Other fields of application are production of potassium chlorate, sodium chlorite, potassium perchlorate, sodium perchlorate, uranium production and weed killing.

Do not use sodium chlorate in mixtures for amateur fireworks. Always follow the safety instructions when handling sodium chlorate to prevent accidents which might lead to great damage to humans or property.

Production units

ALBY Sweden	Address:	Eka Chemicals AB SE-841 44 Alby	STOCKVIK Sweden	Address:	Eka Chemicals AB Box 13000 SE-850 13 Sundsvall
	Fax:	+46 69 01 54 35			
	Phone:	+46 69 01 54 00		Fax:	+46 60 56 93 82
				Phone:	+46 60 13 40 00
AMBES France	Address:	Eka Chimie S.A. Z.I. du Bec FR-33810 Ambès	OULU Finland	Address:	Eka Chemicals OY Box 198 FI-90101 Oulu
	Fax:	+33 556 77 05 08		Fax:	+358 8 3183 3154
	Phone:	+33 556 77 31 40		Phone:	+358 8 3183 3111
MO I RANA Norway	Address:	Eka Chemicals Rana A/S Postboks 17 NO-8601 Mo i Rana			
	Fax:	+47 75 12 69 01			
	Phone:	+47 75 12 69 00			

Safety data sheet co-ordinator
Margaretha Karlsson, phone: +46 87 43 40 00
Approved by PhD Åke Brodén, Eka Chemicals AB, Stockholm

6.1.2.1 *Section 1 – Identification of the substance/preparation and company* This section provides the name, address and telephone number of the supplier along with the product name. One of the first things that should be undertaken upon receipt of the product and MSDS is to check carefully that the product name given on the MSDS matches the label of the drum/container in which the chemical or preparation is received. The issue and revision date of the MSDS should also be given here. The version used should be the most recent. If in doubt, the user should check with the supplier. Suppliers of hazardous chemicals are usually required to provide customers with an updated MSDS with the next shipment of chemical should an important change occur to the original specifications. An example of an important change would be new toxicological information that changes the classification, etc.

Many suppliers will usually re-issue the updated version to all current customers as soon as they can. Again, different countries may have different requirements and it is always best to confirm the relevant regulations.

6.1.2.2 *Section 2 – Composition/information on ingredients* This section provides the identity and quantity of ingredients (if the shipment is a preparation or mixture) or impurities (if a chemical substance) that contribute to the hazardous nature of the product and its ultimate classification. In this context, the term 'hazardous nature' is defined by specific criteria laid out in the legislation of different countries. For example, in the EU the term is as defined by the Dangerous Substances Directive 67/548/EEC, whereas in Canada the equivalent would be as per the Controlled Products Regulations (refer to Chapter 3).

The CAS (Chemical Abstract Service) number should also be given where possible for each of the hazardous ingredients or impurities which are mentioned. CAS numbers are unique for each chemical substance, and as such are useful because it is often easier to refer to a CAS number in discussions rather than to the actual chemical name. The reason for this is that some chemicals have different names which can make matters very confusing. For example, methylbenzene is the official name for toluene, and caustic soda is the trivial name for sodium hydroxide. CAS numbers are very useful, especially if further research requiring database searches etc. is necessary.

Different parts of the world can have slightly different requirements as to what information should be revealed in the MSDS. Therefore, it is always a good idea to check the regulations applicable in the country where the chemical is to be used and confirm that the MSDS fulfils them.

6.1.2.3 *Section 3 – Hazards identification* The purpose of this section is to provide a summary of the most important health, physical and environmental hazards associated with the product and the resulting effects associated with foreseeable use and misuse. Chronic and acute effects, irritancy and likely exposure routes are amongst those items typically mentioned in this section, together with any symptoms arising from exposure to the chemical.

An emergency overview, which highlights the most important and immediate hazards, along with a description of the hazard's appearance can also be given here. This is of use in the event of some kind of emergency because it enables the most significant hazards to be seen at a glance, rather than having to sift through the whole MSDS for important information.

The information provided in this section should reflect what is written in Sections 9, 10, 11 and 12. This is because any hazardous properties of the chemical should be described more fully in these subsequent sections. It is always useful to confirm that this is the situation. If not, the supplier should be contacted.

A word of warning. Different suppliers can present the information in different ways. For example, some may decide to describe the hazards associated during normal exposures to the product, whilst others may provide information related to the worst-case scenario. Therefore, caution should be taken when interpreting information contained in these sections of the MSDS.

6.1.2.4 *Section 4 – First-aid measures* This section provides information on the measures that should immediately be taken if accidental exposure to the product occurs. The recommendations made in this section are given with the aim of minimising the risk of a superficial injury becoming more serious and long-lasting. In some situations the administering of first aid could mean the difference between life and death. What is of vital importance is that prior to working with any chemical, all employees should be made fully aware of the first-aid procedures to use should accidental chemical exposure occur. This includes the location of any emergency showers, eye-wash stations, etc.

6.1.2.5 *Section 5 – Fire-fighting measures* This section gives information needed in the event of a fire, along with a description of the type of fire and explosive properties that the chemical may have. Typical subjects covered include appropriate extinguishing media and very importantly those to be avoided. Autoflammability, flashpoints, explosive properties, and oxidising properties can also be mentioned in this section. However, it should be noted that some suppliers mention flashpoints, oxidising and autoflammability properties in section 9 (Physical and chemical properties) of their MSDS.

Any hazardous decomposition products that may arise during flammable conditions, i.e. acrid or toxic fumes, should also be described. This kind of information is of vital importance so that fire-fighters can take appropriate protective measures and avoid exposure to any hazardous fumes and decomposition products.

During the preparation of the MSDS consideration is given to aspects mentioned in this section when composing the recommendations for handling and storage and accidental release clean-up measures.

6.1.2.6 *Section 6 – Accidental release measures* Specific actions needed in the event of an accidental release or spillage of the product in order to minimise any risk to the environment are described in this section. Particular attention should be given to any personal and environmental precautions that are required. Specific information on what can and cannot be used during the clean-up operation should also be mentioned here. The information provided in this section is based in part on any findings reported in the ecological information section of the MSDS.

6.1.2.7 *Section 7 – Handling and storage*   General precautions needed for the safe handling and storage of the chemical product in order to prevent exposure occurring should be given here. Again, the aim of this section is to provide information relevant to minimising any exposure risk to humans or the environment during normal handling and storage (in other words, to *prevent* any situations that could give rise to a hazard). The handling and storage recommendations given will be influenced by the intrinsic properties of the chemical such as physical and chemical properties, toxicological and ecotoxicological aspects, physical hazards and stability/reactivity.

6.1.2.8 *Section 8 – Exposure controls, personal protection*   This section provides information applicable to maintaining worker exposure to the chemical at the absolute *minimum*. Details of the level of engineering controls needed to reduce (or ideally) minimise exposures to an acceptable level should be mentioned.

Information provided in this section should be based on the known toxicological and physico-chemical properties of the chemical and therefore should make sense in light of what is written in those particular sections. Relevant occupational exposure limits – the PEL of OSHA, TLV of ACGIH, German MAK values, etc. should also be mentioned in this section (see Chapter 4).

Some MSDS list a wide range of exposure limits for different countries, so it is very important to use the correct value. *Always* check the quoted figures with the most recent list of exposure standards for the country where the chemical is being used. It could be the case that an old figure (or even an incorrect value) has been quoted on the MSDS. Never take any of these values for granted.

Recommendations on the best type of personal protection equipment to be used should also be given here. These include details of the specific type of glove (PVC, Neoprene, etc.), work clothes, safety glasses/visor and respiratory apparatus. Information on the specific material providing the best protection should always be mentioned. This is because there is no one material type that can act as a barrier against all chemicals. Thus, should this information not be provided, it is very important to contact the supplier. Logic dictates that if the supplier manufactures the chemical, then it must be protecting its own workers in some way! Otherwise, a little research will be required based on what is known about the chemical nature of the product being handled.

Although general recommendations can be made in an MSDS with regard to respirators, it is always a good idea to obtain advice from a qualified person who can carry out an assessment actually in the workplace.

6.1.2.9 *Section 9 – Physical and chemical properties*   This section is not meant to replace a technical specification despite popular contrary misconception.

The most usual information provided here includes details on the appearance and odour, boiling point/melting points, pH, water solubility and evaporation rate, etc. of the product. Obviously, the information given in this section can be a very useful indication of any potential health hazards that may occur during the handling, storage and use of the product. For example, chemicals with a low boiling point, or high vapour pressure and volatility are likely to create an inhalation risk because they will rapidly vaporise into the atmosphere. Similarly, the pH of the chemical will give an indication of potential irritancy or corrosivity hazards to skin. Therefore, this kind of information is of help when deciding the best handling and storage procedures to follow and personal protection procedures to use.

Another useful piece of information from this section is the description of the product's appearance. By simply reading the description and comparing it to the chemical product, it is possible to see whether or not the MSDS provided relates to the product received. This sounds rather basic, but it is not unknown for the wrong chemical to be shipped out accidentally to a customer. Also, chemicals have a shelf-life, and when this exceeds the specified date, the fact can be reflected by visual changes in appearance.

6.1.2.10 *Section 10 – Stability and reactivity*   The purpose of this section is to inform the user of any circumstances where the chemical may become unstable or react with other materials. Chemical incompatibilities, hazardous reactions, decomposition products and conditions to avoid are some of the sub-headings often seen in this section. Access to this information makes it possible to handle the chemical in a safer manner and contributes to the handling and storage recommendations described in Section 6.1.2.7.

6.1.2.11 *Section 11 – Toxicology*   This section provides information on any tests carried out on animals, and in some cases on humans (e.g. patch testing). The information could be related to the product as a whole, or to its components, or both, and therefore, section 2 on *Composition/information on ingredients* would be taken into account when compiling this section.

Information related to any toxicological health effects arising from exposure should also be given here and ideally this will come from epidemiological studies. However, there is almost always a lack of epidemiological information for chemicals commonly handled in the workplace. Therefore, reliance has to be placed on well-designed animal studies, with any interpretations being made by toxicologists. Typical symptoms arising from exposure can also be listed here. Therefore, do not be surprised if some repetition of information given here and in Section 6.1.2.3 (Hazards identification) is observed. This is because section 3 is based on this section.

It is often the case that there are no actual test data available for the product and therefore test information provided in this section might be derived from other sources including published literature, the MSDS of other suppliers and also from QSAR.

QSAR is the 'quantitative structure activity relationship' and this describes the biological activity of the chemical compound in relation to its molecular structure. Based on this correlation, predictions can be made about how a structurally analogous chemical may react under the same circumstances. Many suppliers of chemical products will usually employ these other resources prior to commissioning toxicological tests. This is for both economic and ethical reasons.

Typical test information that may be highlighted in this section of the MSDS includes the following:

- Acute toxicity
- Sub-acute and sub-chronic toxicity
- Chronic toxicity
- Carcinogenicity
- Mutagenicity
- Reproductive effects
- Teratogenic effects
- Target organ effects
- Other relevant studies

In the United States, information relating to carcinogenicity, if recognised by the National Toxicology Program (NTP), International Agency for Research on Cancer (IARC) and OSHA should also be provided. For Canada, chemicals are identified as carcinogens if they are listed by the IARC and the ACGIH (American Conference of Governmental Industrial Hygienists).

It is quite frequent that an MSDS is lacking toxicological information and the comment 'No information' is often seen. Most national legislations request that any known and relevant information should be provided if available, but there is no requirement to 'go out and get the information' by actually testing the product if no information is available to the supplier. The problem is that in some instances chemical suppliers will not complete the work necessary to provide even the briefest of toxicological data (although the regulations governing this in some countries are better than in others).

As previously mentioned, this requirement does not just mean performing animal studies with the chemical product as an unbelievable amount of information is available from different parts of the world via the Internet. Some of this could be descriptions of different tests already carried out on the same or a structurally similar product. The advent of the Internet has opened up new opportunities in terms of information dissemination to everyone owning a personal computer and a modem which never before existed. Therefore, there really is no excuse for suppliers not being able to provide at least some basic data or even assumptions based on physical and chemical properties. For example, if dealing with a liquid, it is usually possible to measure the pH value. From this, a prediction can be made of the likely irritation or corrosive effects arising upon exposure.

What is the minimum information that should ideally be provided? This would depend on the chemical in question and its intended use. Ideally, information on the acute toxicity of the chemical and an assessment of eye/skin irritancy and sensitisation effects along with a short-term genetic toxicity test, such as the Ames test, would be useful. This way, the user would have something substantial to consult in order to carry out their own workplace risk assessments. If the hazards are not known, the risk involved cannot be assessed. Therefore, always contact the supplier and insist on all relevant information. If the supplier is reluctant, then buy from elsewhere.

6.1.2.12 *Section 12 – Ecological information*   Information provided in this section enables the user to make an assessment of what might happen should the product be released into the environment. Acute and chronic effects, environmental fate, mobility, persistence and bioaccumulation are all aspects of the chemical product that should (hopefully) be mentioned here. From this information, appropriate measures to be taken in the event of a spill can be determined and documented under section 6. The data from this section can also assist in the choice of the most appropriate disposal methods.

Unfortunately, this is often a section of the MSDS where information can be very scant. Therefore, it is often necessary to first and foremost ask for more information from the supplier and if this request provides nothing more, then a literature survey may be needed. In future this situation will be improved as companies and industry as a whole realise the importance of such information.

6.1.2.13 *Section 13 – Disposal considerations* Usually the information given in this section is fairly general, with the main recommendation being to check with local authorities. This is because nationally, local authorities usually have different laws with respect to the disposal of hazardous chemicals. Therefore, different countries will almost certainly have different restrictions on the use of such chemicals. As an extra precaution, confirmation from local authorities or regulatory bodies before disposing of any chemical product is recommended.

6.1.2.14 *Section 14 – Transport information* Section 14 provides details on the transport classification of the product, either by road, rail, air or sea. The shipping regulations for other countries may also be given here.

6.1.2.15 *Section 15 – Regulatory information* This section provides information that can be used to ensure the regulatory compliance of the chemical product. Different areas of the world have their own particular requirements for this section. In the EU, for example, information on the classification and labelling of the chemical will be given in this section in a manner stipulated by the EU Directives on Dangerous Substances and Preparations. In the United States, the applicable federal and state regulations will be mentioned, including those derived from OSHA, CERCLA, CWA, SARA, etc. The WHMIS classification for Canada can also be provided here. Chemical control inventories such as TSCA, EINECS, MITI, or DSL, can also be listed, which is useful if exporting the product to other countries.

6.1.2.16 *Section 16 – Other information* This section should provide a list of any references that were used in the compilation of the MSDS, along with the date of the sheet's preparation, any revision dates, etc.

6.1.3 *Summary*

(1) Always check that the most recent version of the MSDS is available. Telephone the supplier to confirm this if necessary.
(2) Check that the MSDS is complete. If you are dissatisfied with the information provided, telephone the supplier.

(3) Check that the label on the container of the chemical and MSDS are in agreement. If not, then contact the supplier.

(4) Make sure that all workers have access to the MSDS during their shifts. It will be of little use if it is locked up in an office somewhere in the building. Also, ensure that all those who will work with the chemical are fully aware of all the potential hazards associated with it. Also, they should be familiar with the protective measures to be used so as to avoid exposure and the measures to take in the event of an incident.

(5) The MSDS for the products being handled should use the classification and labelling criteria that are relevant to the regulations being abided by. Therefore, if a product is being shipped from the United States to a customer in Europe, the MSDS should comply with the criteria as described in the EU regulations, and vice versa.

(6) The legal requirements concerning the length of time that an MSDS should be kept active can vary from country to country. As a general rule, it is always wise to retain any old versions of the MSDS on file. This can be done by conventional filing of the MSDS or by using microfiche, etc.

(7) Always read the label!

Further reading

American National Standard for Hazardous Industrial Chemicals – Material Safety Data Sheets Preparation
ANSI Z400.1-1993

Guidance on the Compilation of Safety Data Sheets
Chemical Industries Association, November 1998

Safety Data Sheets for Substances and Preparations Dangerous for Supply, 2nd edn
Approved Code of Practice L62, 1994

Directive 93/112/EC amending 91/155/EC Annex – Guide to the Compilation of Safety Data Sheets
OJ, L314, pp. 38–43, 16 December 1993

6.2   How to deal with toxicological or ecotoxicological data gaps

Unfortunately it is very much the case that information relating to toxicological or ecotoxicological effects is scarce or non existent for industrial chemicals. In fact, this can be seen by simply browsing through the 'Chemical Summary Sheets' that are provided for a number of common chemicals used in the pulp and paper industry. Therefore, the 'million dollar question' which either will or has been asked in the past is:

> *'What should we do when there is no ecotoxicological or toxicological information provided?'*

Two approaches that could be taken to answering this question is to assume that the chemical is:

1. Innocent until proven guilty	**?**	2. Guilty until proven innocent

Ideally, it is always best to err on the side of caution and also to remember that possessing no knowledge that a chemical substance (or certain exposure conditions surrounding it for that matter) can cause a particular adverse health effect *is not the same as* having knowledge that it cannot produce a particular adverse health effect.

In reality, there are other approaches that can be taken, some of which have been described in other chapters. For example, if the chemical had not been tested for carcinogenicity then structure activity relationships (SAR) or genotoxicity testing (the Ames test) could both be used to provide an indication of intrinsic chemical behaviour. A chemical's physical structure could also be of assistance – if it were fibrous and small enough to be inhaled it could pose a problem, etc. The Internet is a 'treasure trove' of information, but great care has to be taken when searching in that the information actually obtained is pertinent and not something spurious. Some useful Internet addresses have been given in the 'Further reading' sections of other chapters in the book as will be done here. By connecting to one of the listed homepages, such as that for OSHA, it is possible to link to other relevant Websites.

If a need exists to search for even more information, the best starting point is often the details provided by the supplier of the chemical. Unfortunately, but understandably in the case of formulated products, the supplier will often only reveal the minimum amount of information legally required. Often this information is not sufficient to use in further searches. One solution to this problem is to instigate a dialogue with the supplier and explain the type of information required and the reason for requesting this. More often than not there are ways in which the supplier can help provide the information needed without having to reveal 'the family secrets'. If this approach does not produce results then a confidentiality agreement can be a good alternative.

Unfortunately, the classification and labelling schemes used by different countries (see Chapter 3) deal with positive evidence. Therefore, only substances with the availability of associated data can be classified. These schemes do not cover data gaps. For example, a chemical may be toxic by a route of exposure which has never been investigated, but because there are no data available, it cannot be classified. The problem then is that the recipient of the chemical will be unaware of the necessary precautions required to avoid exposure.

An awareness of the lack of toxicological data has prompted regulatory authorities to act on this problem, for instance in the introduction of the High Production Volume Programme. This was described more fully in Chapter 5, 'Regulatory affairs'.

6.2.1 *How to deal with differences in toxicological information*

Often, toxicological information from different sources for one particular chemical substance can vary considerably. This is potentially a serious problem with which to deal as without actual supporting test reports, it is very difficult, if not impossible, to draw any sound conclusions as to what information should be to relied upon. Variations in the same tests can be attributed to a number of reasons, some of which include species differences, sex differences, or different procedures. These aspects were described in Chapter 1.

Ideally, assessments should be based on test results that have been completed in accordance with recognised standard test protocols (such as the OECD) and GLP. Unfortunately, in the field of industrial chemicals, this is not always possible and often the only test data available are those where little is known concerning the test protocols used, etc. The only advice that can be given is again to err on the side of caution, request the help of a toxicologist or to consult the local trade association and observe how others in the industry use the information.

6.2.2 *Cationic polyelectrolytes and the problems of testing for their environmental effects*

One of the greatest problems faced by suppliers of cationic polyelectrolytes, such as polacrylamides, polyamines, etc., is determining the effects of these chemicals on the environment. Currently there are no official guidelines for testing the biodegradability of such compounds and the available OECD methods are not suitable because the cationic material is strongly adsorbed to the activated sludge used in the procedures and consequently lost from the test system leading to a result that cannot be interpreted. Furthermore, within the aquatic environment, the toxicity of cationic polymers is highly dependent on the chemistry of the water. It is therefore believed that tests carried out in pure water (which standard OECD-type test protocols require) overestimates the true environmental impact of cationic polymers. All cationic substances are potentially toxic to fish because it is believed that the chemical attaches to the anionic gill resulting in deleterious effects on the respiratory function of the organ. In actual environmental conditions, however, it is believed that cationic polymers actually attach to suspended material present in the aqueous phase and thereby their toxicity is significantly reduced.

Work conducted by Goodrich *et al.* (1991) involved adding humic acid to an aqueous environment containing cationic polymers and investigating the effects of toxicity on aquatic life caused by the cationic polyelectrolytes. The study has shown that there is a decrease of the toxicity and that this varies depending on the nature of the polymer.

6.3 Chemical summary sheets

The pages following the 'Further reading' section show summaries of the health and environmental aspects along with the hazard classification for a number of commonly used chemicals in the pulp and paper industry. The aim here is that the information provided complements that of a supplier's MSDS. The details given are by no means comprehensive and *should not* be used in place of a supplier's MSDS. It should also be noted that the inventory listings and exposure limit values can be updated and it is important to confirm the most recent version of an MSDS. No liability is assumed for the accuracy of this information.

Summaries are provided for the following compounds:

- Acrylamide Monomer (for a comparison with polyacrylamide, see Chapter 1)
- Alkyl Ketene Dimer (AKD) wax dispersion (solvent free)
- Alkenyl Succinic Anhydride
- Aluminium Sulphate (solution)
- Aniline Green Dye
- Anionic Polyurethane (aqueous)
- Azo Dye Anionic – yellow
- Azo Dye Cationic – blue
- Bentonite
- Bronopol-type Biocide
- Calcium Hypochlorite
- Cationic Polyacrylamide
- Cationic Starch
- Chlorine
- Colloidal Silica Sol
- Defoamer
- Fluorescent Whitening Agent (stilbene derivative, aqueous solution)
- Fluorescent Whitening Agent (stilbene derivative, anionic)
- Hydrochloric Acid
- Hydrogen Peroxide
- N-methylisothiazolinone-type Biocide
- Polyaluminium Hydroxide Chloride (PAC)
- Polyamide Amine Epichlorohydrin Resin (20% aqueous)
- Polyamine (50% aqueous)
- Polyethylenimine (modified)
- Rosin Size Dispersion (30% casein-based)
- Sodium Chlorate
- Sodium Dithionite
- Sodium Hydroxide
- Sodium Silicate (37% solution)
- Stearic Acid
- Styrene/Acrylate Copolymer (aqueous dispersion)

Reference

Goodrich, M. S., *et al.* (1991) Acute and long-term toxicity of water-soluble cationic polymers to rainbow trout (*Oncorhynchus mykiss*) and the modification of toxicity by humic acid. *Environmental Toxicology and Chemistry*, **10**, 509–515.

Useful Internet addresses

Europe

Toxicological and Ecotoxicological information
ECETOC (European Centre for Ecotoxicology and Toxicology of Chemicals)
http://db1.nihs.go.jp/ecetoc/

IARC
http://www.iarc.fr/

OECD
http://www.oecd.org

Health and Safety Executive (UK)
http://www.hse.gov.UK

CEFIC classification and labelling
http://www.cefic.org/activities/hse/prod_frm.html

VCI Verband der Chemischen Industrie (Germany)
http://www.chemische-industrie.de/

CEFIC – FCA Homepage
http://fca.cefic.org

USA

OSHA Homepage
http://www.OSHA.gov/

ACGIH Homepage
http://www.acgih.org/

NIOSH Homepage
http://www.cdc.gov/niosh/homepage.html

US National Toxicology Program
http://ntp-server.niehs.nih.gov/main_pages/NTP_ARC_PG.html

ICCA HPV Tracking System
http://www.icca-chem.org/hpv/

Canada

http://www.ccohs.ca/

CHEMICAL PRODUCT NAME

Product name: **Acrylamide Monomer**

CAS number: **79-06-1**

In Europe, this is classified as R45 May cause cancer, R46 May cause heritable genetic damage, R24/25 Toxic in contact with skin and if swallowed, R48/23/24/25 Toxic: danger of serious damage to health by prolonged exposure through inhalation, in contact with skin and if swallowed. Note the difference in the classification between this monomer and a cationic polyacrylamide which is shown later in this section. Refer also to Chapter 1 (Toxicity).

TOXICOLOGICAL TESTS AND POTENTIAL HEALTH EFFECTS

Acute effects

Acute oral toxicity	Reported LD_{50} values (rat) range from 159 to 300 mg kg^{-1}
Acute dermal toxicity	LD_{50} (rabbit) = 1000 mg kg^{-1}, LD_{50} (rat) = 400 mg kg^{-1}. Acrylamide is readily absorbed through skin
Acute inhalation toxicity	No test results available

Irritancy and corrosive effects

Skin irritation/Corrosion	Not irritating to skin (rabbits) when administered in a 10% aqueous solution
Eye irritation/Corrosion	Slightly irritating to eyes (rabbits)
Acute respiratory irritation	However, inhalation of dust can cause irritation
Chemical hypersensitivity	No information

Long-term effects

Subchronic/chronic toxicity and other effects	Acrylamide is a neurotoxin both by oral (animals) and inhalation exposure (animals and humans). Toxic effects are central and peripheral neuropathy causes drowsiness, hallucination, distal numbness and ataxia. Recovery is possible after cessation of exposure
Mutagenicity	Causes chromosomal aberrations, dominant lethality, sister chromatid exchanges and unscheduled DNA synthesis
Carcinogenicity	This has been reported as being probably carcinogenic in humans (NTP, IARC, OSHA, ACGIH, EPA). There is inadequate evidence in humans but sufficient evidence from animal studies
Reproductive toxicity and developmental effects (teratogenicity)	No information found on reproductive/developmental effects in humans

CHEMICAL PRODUCT NAME

Product name: **Alkyl Ketene Dimer (AKD) wax dispersion (solvent free)**
CAS number: **84989-41-3**
Intended use: As a sizing agent in the manufacture of paper and board.

AIR CONTAMINATION LIMIT VALUES

Not established

PHYSICAL AND CHEMICAL PROPERTIES

Form colour and odour   White liquid with a faint waxy odour
Melting point/boiling point   Boiling point is 100°C
Vapour pressure   Similar to water
Solubility   Infinitely dispersible in water
pH   2.5–4.5 (commercial product)
Viscosity   < 150 mPa.s
Partition coefficient (log P_{ow})   No test information available
Oxidising properties   No oxidising properties

ECOTOXICOLOGICAL TESTS AND ENVIRONMENTAL EFFECTS

Aquatic toxicity
Acute toxicity to fish: LC_{50} (*Leusisus idus*, 96 h) > 1000 mg l^{-1}. NOEC ≥ 1000 mg l^{-1} (OECD 203)
Acute toxicity to bacteria (*Pseudomonas putia*): EC10 $= 1000$ mg l^{-1}, $EC_{50} = 3000$ mg l^{-1}

Biodegradability
Readily biodegradable: 71% after 28 days (OECD 301B)

Bioaccumulation potential
Product is readily biodegradable and is therefore unlikely to bioaccumulate

TOXICOLOGICAL TESTS AND POTENTIAL HEALTH EFFECTS

Acute effects

Acute oral toxicity LD_{50} (rat) > 2000 mg l^{-1}

Acute dermal toxicity   No information

Acute inhalation toxicity   No information.

Irritancy and corrosive effects

Skin irritation/Corrosion   The main component, AKD wax has been shown to be non-irritating to skin in tests using rabbits. However, the dispersion may cause slight irritation to the skin owing to its low pH

Eye irritation/Corrosion   The main component, AKD wax has been shown to be non-irritating to skin in tests using rabbits. However, the dispersion may cause slight irritation to the skin owing to its low pH

Acute respiratory irritation   No test information available. However, heated aerosols and mists may cause irritation to the naso-respiratory tract

Chemical hypersensitivity   Tests on the dispersion shows it not to be a sensitiser by skin contact (Magnusson & Kligman, OECD 406)

Long-term effects

Subchronic/chronic toxicity and other effects   No test information. However, there has been no known documented cases of adverse effects arising from long-term exposure

Mutagenicity	Tests on the main component, AKD wax, showed no mutagenic activity in the Ames test using five different strains of *Salmonella typhimurium* both with and without the use of metabolic activator
Carcinogenicity	No test information
Reproductive toxicity and developmental effects (teratogenicity)	No test information available

REGULATORY INFORMATION (USA, Canada, Europe)

EU classification and labelling

EU classification	Not classified as dangerous
Risk phrases	None
Safety phrases	None

OSHA hazard communication evaluation (USA)

Does not meet the criteria of 29 CFR 1910.1200

WHMIS classification (Canada)

Does not meet the criteria of a Controlled Product as prescribed by the Controlled Products Regulations

Chemical inventory listings (Y = YES, N = NO)

TSCA	Y	Japan (MITI)	Y
EINECS	Y	DSL (Canada)	Y
AICS	Y		

REFERENCES USED IN THE COMPILATION OF THIS DOCUMENT

HEDSET Data Sheet
2-oxetanone, 3-(C12-C16)-alkyl, 4-(C13-C17)-alkylidene derivatives
CAS number 84989-41-3
May 1999

Various suppliers' MSDS for alkyl ketene dimer wax dispersions

CHEMICAL PRODUCT NAME

Product name: **Alkenyl Succinic Anhydride**
CAS number: **68784-12-3**
Intended use: Sizing for paper and board industry

AIR CONTAMINATION LIMIT VALUES

Not established

PHYSICAL AND CHEMICAL PROPERTIES

Form colour and odour	Clear red-brown viscous liquid with a slight odour
Melting point/boiling point	Boiling point *c.* 100°C
Vapour pressure	< 2500 Pa at 20°C
Solubility	< 1% in water
pH	Hydrolyses in water to give a weakly acidic solution
Viscosity	Approximately 300 mPa.s
Partition coefficient (log P_{ow})	Not known
Oxidising properties	No oxidising properties

ECOTOXICOLOGICAL TESTS AND ENVIRONMENTAL EFFECTS

Aquatic toxicity
Acute toxicity to fish: LC_{50} (*Leuciscius idus*, 96 h) > 10 mg l^{-1} (OECD 203)
Acute toxicity to bacteria: EC10 (*Pseudomonas putia*) > 10 mg l^{-1}, EC_{50} > 8 mg l^{-1}
(DIN 38412-L8 test method)

Biodegradability
Not readily biodegradable (OECD 301B)

Bioaccumulation potential
No information.

TOXICOLOGICAL TESTS AND POTENTIAL HEALTH EFFECTS

Acute effects

Acute oral toxicity	LD_{50} (rat) > 5000 mg kg^{-1} (OECD 401)
Acute dermal toxicity	LD_{50} (dermal, rat) > 5000 mg kg^{-1}
Acute inhalation toxicity	No test information available

Irritancy and corrosive effects

Skin irritation/Corrosion	May cause irritation to the skin. Note that ASA is a powerful degreasing agent (will dry skin) upon prolonged or repeated contact
Eye irritation/Corrosion	May cause irritation to the eyes
Acute respiratory irritation	Inhalation of heated aerosols or mists may irritate the mucous membranes of the naso-respiratory tract
Chemical hypersensitivity	May cause sensitisation by skin contact

Long-term effects

Subchronic/chronic toxicity and other effects	No test information available
Mutagenicity	No test information available
Carcinogenicity	Not listed by IARC, OSHA or ACGIH
Reproductive toxicity and developmental effects (teratogenicity)	No information available

REGULATORY INFORMATION (USA, Canada, Europe)

EU classification and labelling

EU classification	Xi; Irritant
Risk phrases	R43 May cause sensitisation by skin contact
Safety phrases	S28 After contact with skin, wash immediately with plenty of soap and water. S36/37 Wear suitable protective clothing and gloves

OSHA hazard communication evaluation (USA)

Meets the criteria for hazardous material as defined by 29 CFR 1910.1200

WHMIS classification (Canada)

Class D–Poisonous and infectious material
Division 2, sub-division B. Toxic material

Chemical Inventory Listings (Y = YES, N = NO)

TSCA	Y	**MOE (Korea)**	N	**Japan (MITI)**	N
EINECS	Y	**DSL (Canada)**	N		
AICS	N				

REFERENCES USED IN THE COMPILATION OF THIS DOCUMENT

Various suppliers' MSDS

CHEMICAL PRODUCT NAME

Product name:	Aluminium Sulphate (solution)
CAS number:	10043-01-3

AIR CONTAMINATION LIMIT VALUES

UK EH40/98: OES $2\,mg\,m^{-3}$ (LTEL, 8 h TWA) Aluminium salts, soluble
US NIOSH REL-TWA: $2\,mg\,m{-}3$, aluminium salts (soluble salts and alkyls as Al), 1997

PHYSICAL AND CHEMICAL PROPERTIES

Form colour and odour	Clear, colourless and odourless liquid
Melting point/boiling point	Boiling point $c.$ 100°C
Vapour pressure	2000 Pa at 20°C
Solubility	Fully miscible with water
pH	2.5 (commercial product)
Viscosity	$< 25\,mPa.s$ at 20°C
Partition coefficient (log P_{ow})	Not applicable for inorganic compound which dissociates
Oxidising properties	No oxidising properties

ECOTOXICOLOGICAL TESTS AND ENVIRONMENTAL EFFECTS

Aquatic toxicity
Acute toxicity to fish: LC_{50} (*Gambusia affinis*, 96 h) = $37\,mg\,l^{-1}$
Acute toxicity to aquatic invertebrates: EC_{50} (*Daphnia magna*, 15 min) = $136\,mg\,l^{-1}$
The toxicity of aluminium depends on its solubility and speciation both of which are controlled primarily by the pH of the environment

Biodegradability
Not relevant for an inorganic compound

Bioaccumulation potential
Does not bioaccumulate (log P_{ow} is not applicable for an inorganic compound which dissociates)

Mobility and fate
Transport between soil and water will depend on the acidity of the soil (refer to HEDSET)

TOXICOLOGICAL TESTS AND POTENTIAL HEALTH EFFECTS

Acute effects

Acute oral toxicity	LD_{50} (rat) > $5000\,mg\,kg^{-1}$ (OECD 401, GLP)
Acute dermal toxicity	No information
Acute inhalation toxicity	No information.

Irritancy and corrosive effects

Skin irritation/Corrosion	Not irritating (rabbit, OECD 404) using $Al_2(SO_4)3.14.3H_2O$
Eye irritation/Corrosion	Moderately irritating (rabbit, OECD 405) using $Al_2(SO_4)3.14.3H_2O$. However, test using 27 wt% solution gave a result of 'slightly irritating'
Acute respiratory irritation	No test information available. However, aerosols and mists may cause irritation to the respiratory tract
Chemical hypersensitivity	No test information available

Long-term effects

Subchronic/chronic toxicity and other effects	No information available. However there are no known cases of adverse effects arising from long-term exposure to low levels
Mutagenicity	Short-term *in vitro* bacterial tests gave negative results. Not expected to be a mutagenic hazard[1]
Carcinogenicity	Not listed by NTP, IARC or OSHA. Not expected to be a carcinogenic hazard[1]
Reproductive toxicity and developmental effects (teratogenicity)	No known effects. Not expected to be a teratogenic hazard[1]

REGULATORY INFORMATION (USA, Canada, Europe)

EU Classification and labelling

EU classification	Not classified as dangerous
Risk phrases	None
Safety phrases	None

OSHA hazard communication evaluation (USA)

Does not meet the criteria of 29 CFR 1910.1200

WHMIS Classification (Canada)

Does not meet the criteria of a Controlled Product as prescribed by the Controlled Products Regulations

Chemical Inventory Listings (Y = YES, N = NO)

TSCA	Y	MOE (Korea)	Y	Japan (MITI)	Y
EINECS	Y	DSL (Canada)	Y		
AICS	Y				

REFERENCES USED IN THE COMPILATION OF THIS DOCUMENT

1. Leonard, A. & Gerber, G. B. (1988) Mutagenicity, carcinogenicity and teratogenicity of aluminium. *Mutation Research*, **196**, 247–257

HEDSET Data Sheet
Sulphuric acid, aluminium salt
CAS number 10043-01-3
Alcan Chemicals Limited
19/01/95

IUCLID 'Aluminium Sulphate'
ECB – Existing Chemicals
CAS number 10043-01-3
23 October 1995

EKA Chemicals AB
'Aluminium Sulphate' Safety Data Sheet
Revision date July 1998

CHEMICAL PRODUCT NAME

Product name: **Aniline Green Dye**
CAS number: **Aniline Green: 569-64-2**
Intended use: Dyeing for coloured-paper making

AIR CONTAMINATION LIMIT VALUES

Not established

PHYSICAL AND CHEMICAL PROPERTIES

Form colour and odour	Dark green aqueous solution with characteristic odour
Melting point/boiling point	Boiling point $c.$ 100°C
Vapour pressure	No information
Solubility	Miscible in water
pH	2 (10 g l^{-1})
Viscosity	No information
Partition coefficient (log P_{ow})	No information
Oxidising properties	No oxidising properties

ECOTOXICOLOGICAL TESTS AND ENVIRONMENTAL EFFECTS

Aquatic toxicity
Acute toxicity to fish: LC_{50} (*Leuciscius idus*, 48 h) < 1 mg l^{-1}, LC_{50} (Bluegill sunfish, 96 h) = 0.33 mg l^{-1}
Acute toxicity to bacteria: EC_{50} = 50–60 mg l^{-1} (calculated)
Toxicity to aquatic plants (algae): EC_{50} (*Selenastrum capricornutum*, 7 days) < 1 mg l^{-1}

Biodegradability
Less than 50% biodegradation

Bioaccumulation potential
No information

TOXICOLOGICAL TESTS AND POTENTIAL HEALTH EFFECTS

Acute effects

Acute oral toxicity	LD_{50} (mouse) calculated to be $c.$ 950 mg kg^{-1}
Acute dermal toxicity	Harmful by skin contact
Acute inhalation toxicity	No test information available

Irritancy and corrosive effects

Skin irritation/Corrosion	Irritating
Eye irritation/Corrosion	Severely irritating
Acute respiratory irritation	Inhalation of aerosols or mists will irritate the mucous membranes of the naso-respiratory tract
Chemical hypersensitivity	No test information available

Long-term effects

Subchronic/chronic toxicity and other effects	No test information available
Mutagenicity	No test information available
Carcinogenicity	No test information available
Reproductive toxicity and developmental effects (teratogenicity)	No test information available

REGULATORY INFORMATION (USA, Canada, Europe)

EU classification and labelling

EU classification	Xn, Harmful
Risk phrases	R21/22 Harmful in contact with skin and if swallowed. R38 Irritating to skin. R41 Risk of serious damage to eyes
Safety phrases	S26 In case of contact with eyes, rinse immediately with plenty of water and seek medical advice. S28 After contact with skin, wash immediately with plenty of soap and water. S36/37/39 Wear suitable protective clothing, gloves and eye/face protection

OSHA hazard communication evaluation (USA)

Meets the criteria for hazardous material as defined by 29 CFR 1910.1200

WHMIS classification (Canada)

Class D – Poisonous and infectious material
Division 2: Materials causing other toxic effects
Sub-division B: Toxic material

REFERENCES USED IN THE COMPILATION OF THIS DOCUMENT

Various suppliers' MSDS

Chemical substances CD version 5.1 (1991–99) Arbetarskyddsnämnden 'Aniline Green'

CHEMICAL PRODUCT NAME

Product name: **Anionic Polyurethane (aqueous)**
Intended use: Synthetic surface sizing for paper industry

AIR CONTAMINATION LIMIT VALUES

Not established

PHYSICAL AND CHEMICAL PROPERTIES

Form colour and odour	Pale opalescent liquid with characteristic odour
Melting point/boiling point	Boiling point *c.* 100°C
Vapour pressure	< 2500 Pa at 20°C
Solubility	Miscible in water. Polyurethanes are insoluble in water, alcohols and hydrocarbons
pH	7.5–8.2 (commercial product)
Viscosity	15–20 mPa.s at 20°C
Partition coefficient (log P_{ow})	No information
Oxidising properties	No oxidising properties

ECOTOXICOLOGICAL TESTS AND ENVIRONMENTAL EFFECTS

Aquatic toxicity
Acute toxicity to fish: LC_{50} (*Brachydanio rerio*, 96 h) > 1000 mg l^{-1} (OECD 203)
Acute toxicity to bacteria: EC10 (Robra test, *Pseudomonas putia*) = 3800 mg l^{-1}

Biodegradability
Not readily biodegradable (Modified Sturm test)

Bioaccumulation potential
Not expected to occur

TOXICOLOGICAL TESTS AND POTENTIAL HEALTH EFFECTS

<u>Acute effects</u>

Acute oral toxicity	LD_{50} (rat) > 5000 mg kg^{-1} (OECD 401)
Acute dermal toxicity	No test results available
Acute inhalation toxicity	No test results available

<u>Irritancy and corrosive effects</u>

Skin irritation/Corrosion	No test results available. However, may cause very mild but transient irritation to the skin
Eye irritation/Corrosion	No test results available. May cause mild but transient irritation to the eyes
Acute respiratory irritation	No test results available. However, heated aerosols or mists may cause mild irritation
Chemical hypersensitivity	No test results available

Long-term effects

Subchronic/chronic toxicity and other effects	Feeding study in rats gave no mortalities or significant findings at the end of a 90-day period. Polyurethane-covered implants have been used for quite some time in medicine. As a consequence, there is quite an extensive database on polyurethane toxicology
Mutagenicity	Negative results in short-term *in vitro* Ames test (OECD 471)
Carcinogenicity	No test information available
Reproductive toxicity and developmental effects (teratogenicity)	No information

REGULATORY INFORMATION (USA, Canada, Europe)

EU classification and labelling

EU classification	Not classified
Risk phrases	None
Safety phrases	None

OSHA hazard communication evaluation (USA)

Does not meet the criteria for hazardous material as defined by 29 CFR 1910.1200

WHMIS classification (Canada)

Does not meet the criteria of a Controlled Product as prescribed by the Controlled Products Regulations

Chemical inventory listings (Y = YES, N = NO)

TSCA	Y	MOE (Korea)	N	Japan (MITI)	N
EINECS	Y	DSL (Canada)	Y		
AICS	Y				

REFERENCES USED IN THE COMPILATION OF THIS DOCUMENT

Various suppliers' MSDS

CHEMICAL PRODUCT NAME

Product name: **Azo Dye Anionic – yellow**
Intended use: Dyeing in coloured-paper making

AIR CONTAMINATION LIMIT VALUES

Not established

PHYSICAL AND CHEMICAL PROPERTIES

Form colour and odour   Yellow aqueous solution
Melting point/boiling point   Boiling point *c.* 100°C
Solubility   Miscible in water
pH   7.5–8 (20°C)
Partition coefficient (log P_{ow})   No information
Oxidising properties   No oxidising properties

ECOTOXICOLOGICAL TESTS AND ENVIRONMENTAL EFFECTS

Aquatic toxicity
Acute toxicity to fish: LC_{50} (*Oncorhynchus mykiss*, 96 h) $> 100\,mg\,l^{-1}$ (OECD 203)
Acute toxicity to bacteria: $IC_{50} > 100\,mg\,l^{-1}$ (activated sludge, OECD 209)

Biodegradability
Approximately 45% (28 days, DOC; OECD 301E)

Bioaccumulation potential
No information

TOXICOLOGICAL TESTS AND POTENTIAL HEALTH EFFECTS

Acute effects

Acute oral toxicity LD_{50} (rat) $> 2000\,mg\,kg^{-1}$ (88/379/EEC)

Acute dermal toxicity   No test results available

Acute inhalation toxicity   No test results available

Irritancy and corrosive effects

Skin irritation/Corrosion   Non-irritant (rabbit, 88/379/EEC)

Eye irritation/Corrosion   Non-irritant (rabbit, 88/379/EEC)

Acute respiratory irritation No test results available

Chemical hypersensitivity   Not a sensitiser by skin contact (88/379/EEC)

Long-term effects

Subchronic/chronic toxicity No test information available
and other effects

Mutagenicity   No test information available

Carcinogenicity   No test information available

Reproductive toxicity and   No information
developmental effects
(teratogenicity)

REGULATORY INFORMATION (USA, Canada, Europe)

EU classification and labelling

EU classification	Not classified
Risk phrases	None
Safety phrases	None

OSHA hazard communication evaluation (USA)

Does not meet the criteria for hazardous material as defined by 29 CFR 1910.1200

WHMIS classification (Canada)

Does not meet the criteria of a Controlled Product as prescribed by the Controlled Products Regulations

REFERENCES USED IN THE COMPILATION OF THIS DOCUMENT

Various suppliers' MSDS

CHEMICAL PRODUCT NAME

Product name: **Azo Dye Cationic – blue**
Intended use: Dyeing in coloured-paper making

AIR CONTAMINATION LIMIT VALUES

Not established

PHYSICAL AND CHEMICAL PROPERTIES

Form colour and odour	Blue aqueous solution
Melting point/boiling point	Boiling point ca 100°C
Solubility	Miscible in water
pH	Approximately 3 (20°C)
Partition coefficient (log P_{ow})	No information
Oxidising properties	No oxidising properties

ECOTOXICOLOGICAL TESTS AND ENVIRONMENTAL EFFECTS

Aquatic toxicity
Acute toxicity to fish: LC_{50} (*Poecilia reticulata*, 96 h) = 189 mg l^{-1} (OECD 203)
Acute toxicity to bacteria: IC_{50} > 100 mg l^{-1} (activated sludge, OECD 209)

Biodegradability
Greater than 80% (28 days, DOC, OECD 302B)

Bioaccumulation potential
No information

TOXICOLOGICAL TESTS AND POTENTIAL HEALTH EFFECTS

<u>**Acute effects**</u>

Acute oral toxicity	LD_{50} (rat) > 2000 mg kg^{-1} (88/379/EEC)
Acute dermal toxicity	No test results available
Acute inhalation toxicity	No test results available

<u>**Irritancy and corrosive effects**</u>

Skin irritation/Corrosion	Non-irritant (rabbit, 88/379/EEC)
Eye irritation/Corrosion	Non-irritant (rabbit, 88/379/EEC)
Acute respiratory irritation	No test results available. However, inhalation of aerosols or mists may cause some irritation owing to the low pH
Chemical hypersensitivity	Not a sensitiser by skin contact (88/379/EEC)

<u>**Long-term effects**</u>

Subchronic/chronic toxicity and other effects	No test information available
Mutagenicity	No test information available
Carcinogenicity	No test information available
Reproductive toxicity and developmental effects (teratogenicity)	No information

REGULATORY INFORMATION (USA, Canada, Europe)

EU classification and labelling

EU classification	Not classified
Risk phrases	None
Safety phrases	None

OSHA hazard communication evaluation (USA)

Does not meet the criteria for hazardous material as defined by 29 CFR 1910.1200

WHMIS Classification (Canada)

Does not meet the criteria of a Controlled Product as prescribed by the Controlled Products Regulations

REFERENCES USED IN THE COMPILATION OF THIS DOCUMENT

Various suppliers' MSDS

CHEMICAL PRODUCT NAME

Product name: **Bentonite**
CAS number: **14808-60-7**
Intended use: **Adjuvant for drainage and retention aid**

AIR CONTAMINATION LIMIT VALUES

Not established

PHYSICAL AND CHEMICAL PROPERTIES

Form colour and odour   Grey/beige powder
Melting point/boiling point   Melting point > 1000°C
Vapour pressure   Not applicable
Solubility   Insoluble in water
pH   9–11 (at 50 g l^{-1} aqueous suspension)
Viscosity   Not applicable
Partition coefficient (log P_{ow})   No information
Oxidising properties   No oxidising properties

ECOTOXICOLOGICAL TESTS AND ENVIRONMENTAL EFFECTS

Aquatic toxicity
No test information

Biodegradability
Not applicable

Bioaccumulation potential
Not expected to occur

TOXICOLOGICAL TESTS AND POTENTIAL HEALTH EFFECTS

Acute effects

Acute oral toxicity LD_{50} (rat) > 2000 mg kg^{-1}

Acute dermal toxicity   No test results available

Acute inhalation toxicity   No test results available

Irritancy and corrosive effects

Skin irritation/Corrosion   No test results available. However, dust may render skin dry and
                                    chapped upon prolonged or repeated contact

Eye irritation/Corrosion   Moderately irritating to the mucous membranes owing to physical
                                    structure

Acute respiratory irritation   No test results available. However, inhalation of dust can cause
                                    irritation to the mucous membranes

Chemical hypersensitivity   Not senstising

Long-term effects

Sub chronic/chronic toxicity and other effects	No test information available
Mutagenicity	No test information available
Carcinogenicity	No test information available
Reproductive toxicity and developmental effects (teratogenicity)	No information

REGULATORY INFORMATION (USA, Canada, Europe)

EU classification and labelling

EU classification	Not classified
Risk phrases	None
Safety phrases	None

OSHA hazard communication evaluation (USA)

Does not meet the criteria for hazardous material as defined by 29 CFR 1910.1200

WHMIS Classification (Canada)

Does not meet the criteria of a Controlled Product as prescribed by the Controlled Products Regulations

Chemical inventory listings (Y = YES, N = NO)

TSCA	Y	**MOE (Korea)**	Y	**Japan (MITI)**	Y
EINECS	Y	**DSL (Canada)**	Y		
AICS	Y	**Philippines**	Y		

REFERENCES USED IN THE COMPILATION OF THIS DOCUMENT

Various suppliers' MSDS

CHEMICAL PRODUCT NAME

Product name: **Bronopol-type Biocide**
CAS number: **52-51-7 (2-bromo-2-nitro-propandiol)**
Intended use: Biocide

AIR CONTAMINATION LIMIT VALUES

Not established

PHYSICAL AND CHEMICAL PROPERTIES

Form colour and odour   Clear, pale yellow solution
Melting point/boiling point   Boiling point > 100°C
Vapour pressure   Approximately 0.1 mbar (20°C)
Solubility   Completely soluble in water
pH   2.5–4.5 (commercial product)
Viscosity   No information
Partition coefficient (log P_{ow})   0.18
Oxidising properties   No oxidising properties

ECOTOXICOLOGICAL TESTS AND ENVIRONMENTAL EFFECTS

Aquatic toxicity
Acute toxicity to fish: LC_{50} (*Salmo gairdneri*, 96 h) = 412–824 ppm, LC_{50} (*Lepomis machrochirus*, 96 h) = 357 ppm
Acute toxicity to invertebrates: EC_{50} (*Daphnia magna*, 48 h) = 14–28 ppm
Acute toxicity to aquatic plants: EC_{50} (algae, 72 h) = 4–56 ppm

Biodegradability
The active ingredient is biodegradable up to the maximum inhibitory concentration for bacteria

Bioaccumulation potential
Low potential for bioaccumulation

TOXICOLOGICAL TESTS AND POTENTIAL HEALTH EFFECTS

Acute effects

Acute oral toxicity LD_{50} (rat) = 342 mg kg^{-1} (female) and 207 mg kg^{-1} (male)

Acute dermal toxicity LD_{50} (rat) > 1600 mg kg^{-1}

Acute inhalation toxicity   No test information available

Irritancy and corrosive effects

Skin irritation/Corrosion   Contact with the skin can cause pain and irritation

Eye irritation/Corrosion   Irritating (rabbits). Can cause severe pain and irritation

Acute respiratory irritation Inhalation of aerosols or mists may irritate the mucous membranes of the naso-respiratory tract

Chemical hypersensitivity   No skin sensitisation effects known

Long-term effects	
Subchronic/chronic toxicity and other effects	Rats dosed during a 90-day study at 80 mg kg^{-1} daily showed gastrointestinal lesions, respiratory distress and some deaths. A 90-day study using dogs, at 20 mg kg^{-1} daily only produced vomiting effects
Mutagenicity	Short-term *in vitro* genotoxic tests, *Salmonella typhimurium*: negative. V79 cell mutation assay: negative. Short-term *in vivo* genotoxic tests, mouse micronucleus assay: negative. Dominant lethal assay: negative
Carcinogenicity	A two-year toxicity and tumorigenicity test, in which rats received Bronopol at 10, 40 or 60 mg kg^{-1} in drinking water did not affect tumour incidence. Treatment-related non-neoplastic forestomach lesions were attributed to the irritancy of the compound
Reproductive toxicity and developmental effects (teratogenicity)	No known adverse effects

REGULATORY INFORMATION (USA, Canada, Europe)

EU classification and labelling

EU classification	Xn, Harmful
Risk phrases	R22 Harmful if swallowed. R38 Irritating to skin. R41 Risk of serious damage to eyes
Safety phrases	S26 In case of contact with eyes, rinse immediately with plenty of water and seek medical advice. S28 After contact with skin, wash immediately with plenty of soap and water. S36/37/39 Wear suitable protective clothing, gloves and eye/face protection

OSHA hazard communication evaluation (USA)

Meets the criteria for hazardous material as defined by 29 CFR 1910.1200

WHMIS Classification (Canada)

Class D–Poisonous and infectious material
Division 1: Materials causing immediate and serious toxic effects, sub-division B: Toxic material
Division 2: Materials causing other toxic effects, sub-division B: Toxic material

REFERENCES USED IN THE COMPILATION OF THIS DOCUMENT

Various suppliers' MSDS

Medical Library Service
College of Physicians & Surgeons of B.C.
1383 W. 8th Ave.
Vancouver, B.C.
V6H 4C4

CHEMICAL PRODUCT NAME

Product name: **Calcium Hypochlorite**
CAS number: **7778-54-3**
Intended use: Bleaching agent

AIR CONTAMINATION LIMIT VALUES

Not established

PHYSICAL AND CHEMICAL PROPERTIES

Form colour and odour   White/greyish-white powder
Melting point/boiling point   Decomposes above 177°C/350°F
Vapour pressure   Not applicable for solid material
Solubility   Soluble in water; reacts, releasing chlorine gas
pH   11.5 (5% solution)
Viscosity   Not applicable
Partition coefficient (log P_{ow}) No information
Oxidising properties   Strong oxidiser

ECOTOXICOLOGICAL TESTS AND ENVIRONMENTAL EFFECTS

Aquatic toxicity
Acute toxicity to fish: LC_{50} (*Lepomis macrochirus*, 96 h) = 0.049–0.16 mg l^{-1}, LC_{50} (*Oncorhynchus mykiss*, 24 h): *c.* 0.15–0.21 mg l^{-1}
Toxic to algae when exposed to levels of 2 mg l^{-1}

Biodegradability
Not relevant for an inorganic compound

Bioaccumulation potential
Unlikely to bioaccumulate

Mobility
No information available

TOXICOLOGICAL TESTS AND POTENTIAL HEALTH EFFECTS

Acute effects

Acute oral toxicity LD_{50} (rat) = 850 mg kg^{-1}. The material is corrosive and can cause severe burns to the mucous membranes of the mouth, throat and stomach. May cause vomiting and diarrhoea and a sore throat

Acute dermal toxicity LD_{50} (rabbit) > 2000 mg kg^{-1}

Acute inhalation toxicity   No information

Irritancy and corrosive effects

Skin irritation/Corrosion   Calcium hypochlorite dusts are irritating to the skin and in severe cases, corrosive

Eye irritation/Corrosion   Corrosive. Exposure to dust or mists can result in severe effects. Concentrated solutions can cause chemical burns which may result in permanent eye damage

Acute respiratory irritation LC_{50} approximately 1700 mg m^{-3} h^{-1} (rat). Severe irritant to the respiratory tract. Destructive to the mucous membranes and upper respiratory tract. May cause pulmonary oedema. Symptoms of exposure include coughing, shortness of breath, etc.

Chemical hypersensitivity   No information

Long-term effects

Subchronic/chronic toxicity and other effects	No reported cases of adverse effects arising from long-term exposure
Mutagenicity	No *in vivo* data. However calcium hypochlorite tested positive in two short-term, *in vitro* tests (tests are questionable however)
Carcinogenicity	Not known or reported as carcinogenic. Not listed by NTP, IARC or OSHA
Reproductive toxicity and developmental effects (teratogenicity)	No known or reported effects on the reproductive system or foetal development

REGULATORY INFORMATION (USA, Canada, Europe)

EU classification and labelling

EU classification	O, Oxidising. C, Corrosive. N, Very toxic to aquatic organisms
Risk phrases	R8 Contact with combustible material may cause fire. R31 Contact with acids liberates toxic gases. R51 Very toxic to aquatic organisms
Safety phrases	S26 In case of contact with eyes, rinse immediately with plenty of water (for 15 minutes) and seek medical advice. S45 In case of accident, if you feel unwell or experience other effects, seek medical advice immediately (show the label where possible). S61 Avoid release to the environment. Refer to special instructions/MSDS

OSHA hazard communication evaluation (USA)

Meets the criteria for hazardous material as defined by 29 CFR 1910.1200

WHMIS classification (Canada)

Class C – Oxidising material
Class E – Corrosive material

Chemical inventory listings (Y = YES, N = NO)

TSCA	Y	**MOE (Korea)**	Y	**Japan (MITI)**	Y
EINECS	Y	**DSL (Canada)**	Y	**AICS**	Y

REFERENCES USED IN THE COMPILATION OF THIS DOCUMENT

CHEMINFO – CCOHS (97-3) August 1997
Record number: 1322369 'Calcium hypochlorite'
Richardson, M.L. & Gangolli, S., (eds) (1994) *The Dictionary of Substances and their Effects*, Vol. 4, Record Number C32 'Calcium hypochlorite'

Chemical Substances CD version 5.1 (1991–99) Arbetarskyddsnämnden 'Calcium hypochlorite'

IUCLID Data Sheet 'Calcium hypochlorite', ECB Existing Chemicals (23-10-95)

Various suppliers' MSDS

CHEMICAL PRODUCT NAME

Product name: Cationic Polyacrylamide
Intended use: Retention and drainage aid in the paper industry

AIR CONTAMINATION LIMIT VALUES

Not established

PHYSICAL AND CHEMICAL PROPERTIES

Form colour and odour   White powder/granules with a slight odour
Melting point/boiling point   Not applicable
Vapour pressure   Not applicable
Solubility   Up to 1.0 wt% in water
pH   2.5–4.5 (0.5% in water)
Viscosity   Not applicable
Partition coefficient (log *P*ow) No information
Oxidising properties   No oxidising properties

ECOTOXICOLOGICAL TESTS AND ENVIRONMENTAL EFFECTS

Aquatic toxicity
Acute toxicity to fish: LC_{50} (Fathead minnows, 96 h) = 9.33 mg l^{-1}
Acute toxicity to invertebrates: EC_{50} (*Ceriodaphina dubia*, 24h) = 46 mg l^{-1}, EC_{50} (*Ceriodaphina dubia*, 48 h) = 28.2 mg l^{-1}
Acute toxicity to freshwater algae: IC_{50} (*Selenastrum capricornutum*, 96 h) = 636 mg l^{-1}
Results from the 'USEPA Dirty Water test' show that irreversible adsorption onto suspended matter and dissolved organics present in natural waters reduces the availability of the polyacrylamide to aquatic organisms, and therefore its toxicity

Biodegradability
Not readily biodegradable

Bioaccumulation potential
Not expected to occur

Mobility
This product is rapidly eliminated from the aquatic medium through irreversible adsorption onto suspended matter and dissolved organics

TOXICOLOGICAL TESTS AND POTENTIAL HEALTH EFFECTS

<u>**Acute effects**</u>

Acute oral toxicity LD_{50} (rat) > 5000 mg kg^{-1}

Acute dermal toxicity   No test results available

Acute inhalation toxicity   This product is not expected to be toxic by inhalation

<u>**Irritancy and corrosive effects**</u>

Skin irritation/Corrosion   Not irritating to skin (rabbits)

Eye irritation/Corrosion   The product gives no corneal or iridial effects and only slight transient conjunctival effects similar to those produced by all granular materials (Draize test)

Acute respiratory irritation No test results available. However, dust may cause irritation to the mucous membranes of the respiratory tract

Chemical hypersensitivity   Not a sensitiser by skin contact

Long-term effects

Subchronic/chronic toxicity and other effects	A two-year feeding study using rats revealed no adverse health effects. A one-year feeding study in dogs did not reveal any adverse health effects
Mutagenicity	No test information available
Carcinogenicity	No test information available
Reproductive toxicity and developmental effects (teratogenicity)	No information

REGULATORY INFORMATION (USA, Canada, Europe)

EU classification and labelling

EU classification	Not classified
Risk phrases	None
Safety phrases	None

OSHA hazard communication evaluation (USA)

Does not meet the criteria for hazardous material as defined by 29 CFR 1910.1200

WHMIS classification (Canada)

Does not meet the criteria of a Controlled Product as prescribed by the Controlled Products Regulations

Chemical inventory listings (Y = YES, N = NO)

TSCA	**Y**	**MOE (Korea)**	**Y**	**Japan (MITI)**	**Y**
EINECS	**N (polymer)**	**DSL (Canada)**	**Y**		

REFERENCES USED IN THE COMPILATION OF THIS DOCUMENT

Various suppliers' MSDS

CHEMICAL PRODUCT NAME

Product name: **Cationic Starch**
Intended use: Improvement of strength properties in paper

AIR CONTAMINATION LIMIT VALUES

Not established

PHYSICAL AND CHEMICAL PROPERTIES

Form colour and odour	White powder
Melting point/boiling point	Decomposes > 160°C
Vapour pressure	Not applicable
Solubility	Completely in water at > 40°C
pH	6–8 (aqueous)
Viscosity	Not applicable
Partition coefficient (log P_{ow})	No information
Oxidising properties	No oxidising properties

ECOTOXICOLOGICAL TESTS AND ENVIRONMENTAL EFFECTS

Aquatic toxicity
LC_0 (96 h) = 5000 mg l^{-1} (*Bairdiella chrysoura*)

Biodegradability
Good biological degradability: BOD/ThOD = 68.8%

Bioaccumulation potential
Not expected to occur

TOXICOLOGICAL TESTS AND POTENTIAL HEALTH EFFECTS

Acute effects

Acute oral toxicity LD_{50} (rat) > 2000 mg kg^{-1} (OECD 401)

Acute dermal toxicity LD_{50} (rat) > 2000 mg kg^{-1} (OECD 402)

Acute inhalation toxicity   No test results available

Irritancy and corrosive effects

Skin irritation/Corrosion   Not irritating (OECD 404)

Eye irritation/Corrosion   Not irritating (OECD 405)

Acute respiratory irritation   No test results available. However, inhalation of dust can cause irritation to the mucous membranes

Chemical hypersensitivity   No information available

Long-term effects

Subchronic/chronic toxicity and other effects   No test information available

Mutagenicity   No test information available

Carcinogenicity   No test information available

Reproductive toxicity and developmental effects (teratogenicity)   No information

REGULATORY INFORMATION (USA, Canada, Europe)

EU classification and labelling

EU classification   Not classified

Risk phrases   None

Safety phrases   None

OSHA hazard communication evaluation (USA)

Does not meet the criteria for hazardous material as defined by 29 CFR 1910.1200

WHMIS classification (Canada)

Does not meet the criteria of a Controlled Product as prescribed by the Controlled Products Regulations

Chemical inventory listings (Y = YES, N = NO)

TSCA	Y	**MOE (Korea)**	Y	**Japan (MITI)**	Y
EINECS	Y	**DSL (Canada)**	Y	**AICS**	Y

REFERENCES USED IN THE COMPILATION OF THIS DOCUMENT

Various suppliers' MSDS

CHEMICAL PRODUCT NAME

Product name: **Chlorine**
CAS number: **7782-50-5**
Intended use: Pulp bleaching

AIR CONTAMINATION LIMIT VALUES

OSHA PEL: Ceiling limit 1 ppm (3 mg m^{-3})
NIOSH REL: Ceiling limit 0.5 ppm (1.45 mg m^{-3})
ACGIH TLV: 0.5 ppm TWA (1.5 mg m^{-3}); 1 ppm STEL (2.9 mg m^{-3}) A4
OEL UK: 1 ppm TWA (3 mg m^{-3}); 3 ppm STEL (9 mg m^{-3})
OEL Sweden: 0.5 ppm TWA (1.5 mg m^{-3})

PHYSICAL AND CHEMICAL PROPERTIES

Form colour and odour	Greenish-yellow gas with pungent odour
Melting point/boiling point	Boiling point: $-35°C/-30°F$
Vapour pressure	638.4 kPa (20°C)
Solubility	Slightly soluble in water (0.7 g per 100 g at 20°C)
pH	Not applicable
Viscosity	Not applicable
Partition coefficient (log P_{ow})	Not applicable
Oxidising properties	Strong oxidiser

ECOTOXICOLOGICAL TESTS AND ENVIRONMENTAL EFFECTS

Aquatic toxicity
Acute toxicity to fish: LC$_{50}$ (*Salmo gairdneri*, 96 h) = 0.014 mg l^{-1}
Acute toxicity to aquatic invertebrates: EC$_{50}$ (Daphnia, 48 h) = 0.028 mg l^{-1}
Acute toxicity to aquatic plants: IC$_{50}$ (72 h growth inhibition test) = 0.76 mg l^{-1}
Chlorine is very reactive and will form hydrochloric acid in contact with water. Chlorinated organic
   compounds can be formed from organic material in the water

Biodegradability
Not applicable. Note that it has been reported that activated sludge is not affected by concentrations up
   to 10 mg l^{-1}

Bioaccumulation potential
Unlikely to accumulate. It will react with water and tissues

Mobility and environmental fate
Chlorine is a gas and is therefore likely to predominate mainly in the air compartment

TOXICOLOGICAL TESTS AND POTENTIAL HEALTH EFFECTS

Acute effects

Acute oral toxicity	Not an anticipated route of exposure as is a gas at normal atmospheric conditions
Acute dermal toxicity	Not available
Acute inhalation toxicity	LC$_{50}$ (rat, 1 h) = 293 ppm (equivalent to 147 ppm per 4 h exposure), LC$_{50}$ (mouse, 1 h) = 137 ppm

Irritancy and corrosive effects

Skin irritation/Corrosion	High concentrations of gas causes severe irritation with symptoms which can include burning, prickling sensations, reddening and blisters. Direct contact with liquid chlorine causes burns
Eye irritation/Corrosion	Severe eye irritant. Symptoms can include stinging, burning sensations and tears
Acute respiratory irritation	Severe irritant to the nose, throat and upper respiratory tract
Chemical hypersensitivity	No information

Long-term effects

Subchronic/chronic toxicity and other effects	Repeated/prolonged exposure to 5 ppm chlorine gas produces nose inflammation and tooth enamel corrosion
Mutagenicity	Chlorine does not appear to be mutagenic in short-term tests. Ames test: negative (with and without metabolising system)
Carcinogenicity	Not listed by IARC, OSHA or ACGIH. No evidence of carcinogenicity in humans or animals
Reproductive toxicity	Multi-generation study where rats were administered 100 mg l^{-1} chlorine in drinking water over 7 consecutive generations showed no adverse effects on fertility, life-span or growth pattern
Developmental effects (teratogenicity)	No information

REGULATORY INFORMATION (USA, Canada, Europe)

EU classification and labelling

EU classification	T, Toxic; Xi, Irritant; N, R50
Risk phrases	R23 Toxic by inhalation. R36/37/38 Irritating to eyes, respiratory system and skin. R50 Very toxic to aquatic organisms
Safety phrases	S9 Keep container in a well-ventilated place. S45 In case of accident, if you feel unwell or experience other effects, seek medical advice immediately (show the label where possible). S61 Avoid release to the environment. Refer to special instructions/MSDS

OSHA hazard communication evaluation (USA)

Meets the criteria for hazardous material as defined by 29 CFR 1910.1200

WHMIS classification (Canada)

Class A – Compressed gas
Class C – Oxidising material.
Class D1A – Poisonous and infectious material – Immediate and serious effects – Very toxic
Class D2A – Poisonous and infectious material – Other effects – Very toxic
Class E – Corrosive material

Chemical inventory listings (Y = YES, N = NO)

TSCA	Y	MOE (Korea)	Y	AICS	Y
EINECS	Y	DSL (Canada)	Y		

REFERENCES USED IN THE COMPILATION OF THIS DOCUMENT

STN Easy Regulated Chemical Lists
http://stneasy.FIZ-Karlsruhe.de/

CHEMINFO – CCOHS (Issue: 98-2)
Record number 85 'Chlorine'

Richardson, M.L. & Gangolli, S. (eds) (1994) *The Dictionary of Substances and their Effects*, Vol.4,
Record number C127 'Chlorine'

Chemical Substances CD version 5.1 (1991–99) Arbetarskyddsnämnden 'Chlorine'

IUCLID Data Sheet 'Chlorine', ECB Existing Chemicals (1995)
CAS number: 7782-50-5

Fluka Chemicals Corp. USA 'Chlorine 99.5 + %', Catalogue number: 295132, Sigma Aldrich

ChemAdvisor® CCOHS
ChemAdvisor record number: 7833 'Chlorine'; Issue: July 1997

CHEMICAL PRODUCT NAME

Product name:	**Colloidal Silica Sol**
CAS number:	**7631-86-9**
Intended use:	As a retention and drainage agent in the manufacture of paper and board

AIR CONTAMINATION LIMIT VALUES

Amorphous silica	OEL UK: Total inhalable dust LTEL (8 h, TWA) 6 mg m^{-3}. Respirable dust: LTEL (8 h, TWA) 2.4 mg m^{-3} (1998). NIOSH REL: (TWA) 6 mg m^{-3}

PHYSICAL AND CHEMICAL PROPERTIES

Form colour and odour	Clear, colourless and odourless
Melting point/boiling point	Boiling point: 100°C
Vapour pressure	Similar to water
Solubility	Completely soluble in water
pH	9.7–10.5 (commercial product)
Viscosity	< 10 mPa.s (25°C)
Partition coefficient (log P_{ow})	Not applicable
Oxidising properties	No oxidising properties

ECOTOXICOLOGICAL TESTS AND ENVIRONMENTAL EFFECTS

Aquatic toxicity
Acute toxicity to fish: LC$_{50}$ (*Brachydanio rerio*, 96 h) = 7600 mg l^{-1}
Acute toxicity to aquatic plants: IC$_{50}$ (*Selenastrum capricornutum*, 48 h growth inhibition test) = 440 mg l^{-1}, EC10 = 140 mg l^{-1}, LOEC = 120 mg l^{-1}, NOEC = 60 mg l^{-1}

Biodegradability
Not relevant for an inorganic compound

Bioaccumulation potential
No potential for bioaccumulation

Mobility
Amorphous silica is ubiquitous in the environment

TOXICOLOGICAL TESTS AND POTENTIAL HEALTH EFFECTS

Acute effects

Acute oral toxicity	LD$_{50}$ (rat) > 15,000 mg kg^{-1}
Acute dermal toxicity	No test information available
Acute inhalation toxicity	No test information available

Irritancy and corrosive effects

Skin irritation/Corrosion	Mildly irritating (rabbit). Prolonged or repeated contact may cause defatting of the skin which could lead to dermatitis
Eye irritation/Corrosion	Mildly irritating (rabbit)
Acute respiratory irritation	No test information available. However, aerosols and mists may cause irritation to the respiratory tract owing to its high pH
Chemical hypersensitivity	Not a sensitiser by skin contact (Magnusson & Kligman, OECD 406)

Medical Library Service
College of Physicians & Surgeons of B.C.
1383 W. 8th Ave.
Vancouver, B.C.
V6H 4C4

Long-term effects

Subchronic/chronic toxicity and other effects	This product contains amorphous silicon dioxide. No cases of silicosis have been reported from amorphous silica
Mutagenicity	Short-term *in vitro* bacterial 'Ames test' gave negative results
Carcinogenicity	Not listed by NTP, IARC or OSHA
Reproductive toxicity and developmental effects (teratogenicity)	No test information available

REGULATORY INFORMATION (USA, Canada, Europe)

EU classification and labelling

EU classification	Not classified as dangerous
Risk phrases	None
Safety phrases	None

OSHA hazard communication evaluation (USA)

Does not meet the criteria for hazardous material as defined by 29 CFR 1910.1200

WHMIS classification (Canada)

Does not meet the criteria of a Controlled Product as prescribed by the Controlled Products Regulations

Chemical inventory listings (Y = YES, N = NO)

TSCA	Y	**MOE (Korea)**	Y	**Japan (MITI)**	Y
EINECS	Y	**DSL (Canada)**	Y	**AICS**	Y
Philippines	Y				

REFERENCES USED IN THE COMPILATION OF THIS DOCUMENT

EKA Chemicals AB, Bohus, Sweden, Paper Chemicals Division: 'BMA 0' (March 2000)

CHEMICAL PRODUCT NAME

Product name:	**Defoamer**
Chemical type:	**Polydimethylsiloxane emulsion**
Intended use:	Foam prevention

AIR CONTAMINATION LIMIT VALUES

Not established

PHYSICAL AND CHEMICAL PROPERTIES

Form colour and odour	Milky white liquid
Melting point/boiling point	Boiling point $c.$ 100°C
Solubility	Miscible in water
pH	6–8
Partition coefficient (log P_{ow})	No information
Oxidising properties	No oxidising properties

ECOTOXICOLOGICAL TESTS AND ENVIRONMENTAL EFFECTS

Aquatic toxicity
Concentrated defoamer may adversely affect fish life

Biodegradability
No information

Bioaccumulation potential
No information

TOXICOLOGICAL TESTS AND POTENTIAL HEALTH EFFECTS

Acute effects

Acute oral toxicity	No test information. However, product is not expected to be hazardous
Acute dermal toxicity	No test information. However, product is not expected to be hazardous
Acute inhalation toxicity	No test results available

Irritancy and corrosive effects

Skin irritation/Corrosion	Low irritation potential
Eye irritation/Corrosion	Low irritation potential. However, contact may produce an oil film over the eye ball causing a harmless reversible transient dimness of sight
Acute respiratory irritation	No test results available
Chemical hypersensitivity	No information

Long-term effects

Subchronic/chronic toxicity and other effects	No test information available
Mutagenicity	Not mutagenic
Carcinogenicity	Not carcinogenic
Reproductive toxicity and developmental effects (teratogenicity)	No information

REGULATORY INFORMATION (USA, Canada, Europe)

<u>EU classification and labelling</u>

EU classification	Not classified
Risk phrases	None
Safety phrases	None

<u>OSHA hazard communication evaluation (USA)</u>

Does not meet the criteria for hazardous material as defined by 29 CFR 1910.1200

<u>WHMIS classification (Canada)</u>

Does not meet the criteria of a Controlled Product as prescribed by the Controlled Products Regulations

REFERENCES USED IN THE COMPILATION OF THIS DOCUMENT

Various suppliers' MSDS

CHEMICAL PRODUCT NAME

Product name:	**Fluorescent Whitening Agent (stilbene derivative, aqueous solution)**
Intended use:	To make paper appear whiter

AIR CONTAMINATION LIMIT VALUES

Not established

PHYSICAL AND CHEMICAL PROPERTIES

Form colour and odour	Pale yellow liquid
Melting point/boiling point	Boiling point: 100°C (1013 hPa)
Solubility	Miscible in water
pH	10 (20°C)
Partition coefficient (log P_{ow})	No information
Oxidising properties	No oxidising properties

ECOTOXICOLOGICAL TESTS AND ENVIRONMENTAL EFFECTS

Aquatic toxicity
Acute toxicity to fish: LC_{50} (*Oncorhynchus mykiss*, 96 h) > 100 mg l^{-1} (OECD 203)
Acute toxicity to bacteria: IC_{50} > 1000 mg l^{-1}

Biodegradability
17% COD

Bioaccumulation potential
No information

TOXICOLOGICAL TESTS AND POTENTIAL HEALTH EFFECTS

Acute effects

Acute oral toxicity	LD_{50} (rat) > 2000 mg kg^{-1} (OECD 401)
Acute dermal toxicity	No test results available
Acute inhalation toxicity	No test results available

Irritancy and corrosive effects

Skin irritation/Corrosion	Non-irritant (OECD 404)
Eye irritation/Corrosion	Non-irritant (OECD 405)
Acute respiratory irritation	No test results available. However, inhalation of aerosols or mists may cause irritation owing to the high pH
Chemical hypersensitivity	No test results available

Long-term effects

Subchronic/chronic toxicity and other effects	No test information available
Mutagenicity	No test information available
Carcinogenicity	No test information available
Reproductive toxicity and developmental effects (teratogenicity)	No information

REGULATORY INFORMATION (USA, Canada, Europe)

EU classification and labelling

EU classification	Not classified
Risk phrases	None
Safety phrases	None

OSHA hazard communication evaluation (USA)

Does not meet the criteria for hazardous material as defined by 29 CFR 1910.1200

WHMIS classification (Canada)

Does not meet the criteria of a Controlled Product as prescribed by the Controlled Products Regulations

REFERENCES USED IN THE COMPILATION OF THIS DOCUMENT

Various suppliers' MSDS

CHEMICAL PRODUCT NAME

Product name: **Fluorescent Whitening Agent (stilbene derivative, anionic)**
Intended use: To make paper appear whiter

AIR CONTAMINATION LIMIT VALUES

Not established

PHYSICAL AND CHEMICAL PROPERTIES

Form colour and odour	White aqueous solution
Melting point/boiling point	Boiling point: 100°C (1013 hPa)
Solubility	Miscible in water
pH	Approximately 9 (20°C)
Partition coefficient (log P_{ow})	No information
Oxidising properties	No oxidising properties

ECOTOXICOLOGICAL TESTS AND ENVIRONMENTAL EFFECTS

Aquatic toxicity
Acute toxicity to fish: LC_{50} (*Oncorhynchus mykiss*, 96 h) > 1000 mg l^{-1} (OECD 203)
Acute toxicity to bacteria: IC_{50} > 1000 mg l^{-1} (activated sludge, OECD 209)

Biodegradability
Approximately 55% TOC

Bioaccumulation potential
No information

TOXICOLOGICAL TESTS AND POTENTIAL HEALTH EFFECTS

<u>Acute effects</u>

Acute oral toxicity	LD_{50} (rat) > 5000 mg kg^{-1}
Acute dermal toxicity	No test results available
Acute inhalation toxicity	No test results available

<u>Irritancy and corrosive effects</u>

Skin irritation/Corrosion	Non-irritant
Eye irritation/Corrosion	Non-irritant
Acute respiratory irritation	No test results available. However, inhalation of aerosols or mists may cause irritation owing to the high pH
Chemical hypersensitivity	No test results available

<u>Long-term effects</u>

Subchronic/chronic toxicity and other effects	No test information available
Mutagenicity	No test information available
Carcinogenicity	No test information available
Reproductive toxicity and developmental effects (teratogenicity)	No information

REGULATORY INFORMATION (USA, Canada, Europe)

EU classification and labelling

EU classification	Not classified
Risk phrases	None
Safety phrases	None

OSHA hazard communication evaluation (USA)

Does not meet the criteria for hazardous material as defined by 29 CFR 1910.1200

WHMIS classification (Canada)

Does not meet the criteria of a Controlled Product as prescribed by the Controlled Products Regulations

REFERENCES USED IN THE COMPILATION OF THIS DOCUMENT

Various suppliers' MSDS

CHEMICAL PRODUCT NAME

Product name:	**Hydrochloric Acid**
CAS number:	**7647-01-0**
Intended use:	Diverse applications in the pulp and paper industry

AIR CONTAMINATION LIMIT VALUES

ACGIH TLV-C: 5 ppm (7.5 mg m^{-3})
OSHA PEL-C: 5 ppm (7 mg m^{-3})
UK OEL: STEL 5 ppm (7.6 mg m^{-3})

PHYSICAL AND CHEMICAL PROPERTIES

Form colour and odour	Colourless/slightly yellow liquid, choking odour
Melting point/boiling point	Boiling point: 100°C
Vapour pressure	100 mmHg (13.3 kPa) at 20°C
Solubility	Completely in water. Soluble in certain organic solvents, e.g. methanol, ethanol, diethyl ether
pH	1.1 (0.1N)
Viscosity	No information
Partition coefficient (log P_{ow})	No information
Oxidising properties	None

ECOTOXICOLOGICAL TESTS AND ENVIRONMENTAL EFFECTS

Aquatic toxicity
Acute toxicity to fish: LC$_{50}$ (Mosquito fish, 96 h) = 232 mg l^{-1}
Acute toxicity to aquatic invertebrates: EC$_{50}$ (Daphnia, 48 h) > 56 mg l^{-1}
Acute effects are due to the lowered pH and corrosive damage

Biodegradability
Not biodegradable as such

Bioaccumulation potential
Unlikely to accumulate

TOXICOLOGICAL TESTS AND POTENTIAL HEALTH EFFECTS

Acute effects

Acute oral toxicity	LD$_{50}$ (rabbit) = 900 mg kg^{-1}. Ingestion likely to cause pain, burns to the mucous membranes of the throat, mouth, oesophagus and stomach. Other symptoms include gastric pain and nausea
Acute dermal toxicity	No information available
Acute inhalation toxicity	LC$_{50}$ (rat, 1 h) = 3124 ppm; LC$_{50}$ (mouse, 4 h) = 388 mg m^{-3} (aerosol, mist)

Irritancy and corrosive effects

Skin irritation/Corrosion	Corrosive. Repeated or prolonged contact to dilute solutions may cause dermatitis
Eye irritation/Corrosion	Corrosive. Vapours from 37% concentration are immediately irritating. The severity of the damage depends on the quantity, concentration and duration of exposure
Acute respiratory irritation	Irritating. Exposure to gas or fumes may cause choking, burning sensation
Chemical hypersensitivity	No information available

Long-term effects

Subchronic/chronic toxicity and other effects	Repeated skin contact to mist or diluted liquid may cause dermatitis. Repeated inhalation may cause bleeding of gums and nose and chronic bronchitis
Mutagenicity	Questionable positive evidence in short-term *in vitro* tests
Carcinogenicity	No carcinogenic status
Reproductive toxicity and developmental effects (teratogenicity)	No information

REGULATORY INFORMATION (USA, Canada, Europe)

EU classification and labelling

EU classification	C, Corrosive, Xi, Irritant
Risk phrases	R34 Causes burns. R37 Irritating to respiratory system
Safety phrases	S26 In case of contact with eyes, rinse immediately with plenty of water (for 15 minutes) and seek medical advice. S45 In case of accident or if you feel unwell, seek medical advice immediately (show the label where possible)

OSHA hazard communication evaluation (USA)

Meets the criteria for hazardous material as defined by 29 CFR 1910.1200

WHMIS classification (Canada)

Class D1A – Poisonous and infectious material – Immediate and serious effects – Very toxic
Class E – Corrosive material

REFERENCES USED IN THE COMPILATION OF THIS DOCUMENT

STN Easy. MSDS – CCOHS 2000
AN 1992755 'Hydrochloric acid'

CHEMINFO – CCOHS (98-2) May 1998
Record number: 13 'Hydrochloric acid'

STN Easy. Regulated Chemical Lists
'Hydrochloric acid' CAS number: 1344-09-8
http://stneasy.FIZ.Karlsruhe.de/

Chemical substances CD version 5.1 (1991–99) Arbetarskyddsnämnden 'Hydrochloric acid'

CHEMICAL PRODUCT NAME

Product name:	Hydrogen Peroxide
CAS number:	7722-84-1
Intended use:	Pulp bleaching

AIR CONTAMINATION LIMIT VALUES

TLV-TWA:1 ppm (1.4 mg m^{-3}), NIOSH/OSHA TWA: 1 ppm (1.4 mg m^{-3})
ACGIH-TWA: 1 ppm (1.4 mg m^{-3}) A3
OEL UK LTEL: 1 ppm (1.5 mg m^{-3}), STEL: 2 ppm (3 mg m^{-3})

PHYSICAL AND CHEMICAL PROPERTIES

Form colour and odour	Clear, colourless, odourless liquid
Melting point/boiling point	Freezing point: –33°C. Boiling point: 152°C
Vapour pressure	48 Pa (30°C)
Solubility	Miscible in water; organic solvent: diethyl ether
pH	1–3 (concentrated solution)
Viscosity	No information
Partition coefficient (log P_{ow})	No information
Oxidising properties	Strong oxidiser

ECOTOXICOLOGICAL TESTS AND ENVIRONMENTAL EFFECTS

Aquatic toxicity
Acute toxicity to fish: LC$_{50}$ (*Pimephales promelas*, 96 h) = 16.4 mg l^{-1}
Acute toxicity to aquatic invertebrates: EC$_{50}$ (*Daphnia pulex*, 48 h) = 2.4 mg l^{-1}
Acute toxicity to aquatic plants: IC$_{50}$ (*Chlorella vulgaris*, 72 h growth inhibition test) = 2.5 mg l^{-1}

Biodegradability
Not biodegradable as such. However, product will rapidly decompose to water and oxygen. The rate of degradation is dependent on the concentration of minerals and microbiological activity

Bioaccumulation potential
Unlikely to accumulate as it decomposes so rapidly

Mobility and fate
The product is completely miscible in water and therefore will have a high mobility. However, the product will rapidly decompose and the rate of degradation will depend on the concentration of minerals and microbiological activity present in the aqueous environment. In soil, hydrogen peroxide will be rapidly decomposed, but the degradation rate will again depend on the concentration of minerals and microbiological activity

TOXICOLOGICAL TESTS AND POTENTIAL HEALTH EFFECTS

Acute effects

Acute oral toxicity	LD$_{50}$ (rat) = 1000 mg kg^{-1}. Ingestion likely to cause pain, burns to the mucous membranes of the throat, mouth, oesophagus and stomach. Other symptoms include gastric pain and nausea. Once ingested, it rapidly decomposes to oxygen which can cause bleeding and severe internal injuries
Acute dermal toxicity	LD$_{50}$ (rat) = 4060 mg kg^{-1}; LD$_0$ (rabbit) = 2000 mg kg^{-1}
Acute inhalation toxicity	LC$_{50}$ (rat, 4 h) = 2000 mg m^{-3}; LD$_0$ (rabbit) = 2000 mg kg^{-1}. Inhalation is very irritating to the respiratory tract. Symptoms include coughing, dizziness and a sore throat. Inhalation of the aerosol mist causes inflammation and pulmonary oedema

Irritancy and corrosive effects

Skin irritation/Corrosion	Draize test: 50% solution, 4 h exposure (rabbit) produced severe effects. Classified as corrosive. Skin contact immediately causes bleaching of the skin, then redness, blisters and possibly burns
Eye irritation/Corrosion	Draize test: 10% to 12% strength produced severe damage to the eyes which was irreversible when not immediately washed out. Eye contact can cause permanent damage to the cornea
Acute respiratory irritation	Inhalation of hot vapours or mist produces severe irritation to the respiratory airways
Chemical hypersensitivity	Not a sensitiser by skin contact

Long-term effects

Subchronic/chronic toxicity and other effects	Six-month study using dogs who were exposed to an average vapour concentration of 7 ppm for 6 h day^{-1}, 5 days week^{-1} and other effects had lung irritation but no other significant effects
Mutagenicity	Positive results in some short-term *in vitro* tests. Negative results in *in vivo* tests reported
Carcinogenicity	No clear evidence of carcinogenicity from animal studies. Not listed by IARC, OSHA or ACGIH
Reproductive toxicity and developmental effects (teratogenicity)	No information

REGULATORY INFORMATION (USA, Canada, Europe)

EU classification and labelling

EU classification	O, Oxidising, C, Corrosive
Risk phrases	R8 Contact with combustible material may cause fire. R34 Causes burns
Safety phrases	S3 Keep in a cool place. S28 After contact with skin, wash immediately with plenty of water. S36/39 Wear suitable protective clothing and eye/face protection. S45 In case of accident or if you feel unwell, seek medical advice immediately (show the label where possible)

OSHA hazard communication evaluation (USA)

Meets the criteria for hazardous material as defined by 29 CFR 1910.1200

WHMIS classification (Canada)

Class C – Oxidizing material
Class D1B – Poisonous and infectious material immediate and serious effects – Toxic
Class E – Corrosive material

Chemical inventory listings (Y = YES, N = NO)

TSCA	Y	MOE (Korea)	Y	**Japan (MITI)**	Y
EINECS	Y	DSL (Canada)	Y		
AICS	Y	Philippines	Y		

REFERENCES USED IN THE COMPILATION OF THIS DOCUMENT

Chem Advisor® Inc.
Record number: 7751, 'Hydrogen peroxide'
Issue date: 97-3 (July 1997)

STN Easy. MSDS – CCOHS 1999
AN 2027031 number: 54038

CHEMINFO – CCOHS (97-3) August 1997
Record number: 77 'Hydrogen peroxide'

EKA Chemicals AB, Bohus, Sweden, Bleaching Chemicals Division
'Hydrogen peroxide 35 wt%' (1996)

Richardson, M.L. & Gangolli, S., eds. (1994) *The Dictionary of Substances and their Effects*, Vol.4, Record number H96

Chemical Substances CD version 5.1 (1991–99) Arbetarskyddsnämnden 'Hydrogen peroxide'

IUCLID Data Sheet 'Hydrogen peroxide', ECB Existing Chemicals

Fluka Chemicals Corp. USA 'Hydrogen peroxide 30%' Catalogue number: 95321, Sigma Aldrich

CHEMICAL PRODUCT NAME

Product name:	**N-Methylisothiazolinone-type Biocide**
CAS number:	**Mixture of 5-chloro-2-methyl-2H-isothiazol-3-one and 2-methyl-2H-isothiazol-3-one (3:1), CAS numbers: 26172-44-4 and 2682-20-4, respectively**
Intended use:	Biocide

AIR CONTAMINATION LIMIT VALUES

No information

PHYSICAL AND CHEMICAL PROPERTIES

Form colour and odour	Clear, pale yellow liquid with slight odour
Melting point/boiling point	Boiling point = 100°C
Vapour pressure	23 mbar (20°C)
Solubility	Completely in water
pH	4.1 (10% solution)
Viscosity	No information
Partition coefficient (log P_{ow})	No information
Oxidising properties	No oxidising properties

ECOTOXICOLOGICAL TESTS AND ENVIRONMENTAL EFFECTS

Aquatic toxicity
Acute toxicity to fish: LC_{50} (*Oncorhynchus mykiss*, 96 h) = 13.9 mg l^{-1}
Acute toxicity to invertebrates: Immobilisation, EC_{50} (*Daphnia magna*, 48 h) = 7.78 mg l^{-1}

Biodegradability
The active ingredient is biodegradable up to the maximum inhibitory concentration for bacteria. Note:
   product can affect the A_{OX} value of effluent water

Bioaccumulation potential
Low potential for bioaccumulation

TOXICOLOGICAL TESTS AND POTENTIAL HEALTH EFFECTS

Acute effects	
Acute oral toxicity	LD_{50} (rat) > 2000 mg kg^{-1} (OECD 401/EPA 81-1)
Acute dermal toxicity	LD_{50} (rat) > 2000 mg kg^{-1} (OECD 402/EPA 81-2)
Acute inhalation toxicity	LC_{50} (rat, 4 h exposure) = 12.3 mg l^{-1} (OECD 403/EPA 81-3)
Irritancy and corrosive effects	
Skin irritation/Corrosion	Irritating (rabbits) (EPA FIFRA 81-5)
Eye irritation/Corrosion	Moderate to severe irritant (EPA FIFRA 81-4)
Acute respiratory irritation	Inhalation of aerosols or mists will irritate the mucous membranes of the naso-respiratory tract
Chemical hypersensitivity	May cause skin sensitisation (EPA-81-6, Magnusson & Kligman test)
Long-term effects	
Subchronic/chronic toxicity and other effects	A 90-day oral feeding study in dogs gave no organ or systemic toxicity up to 750 ppm as active ingredient
Mutagenicity	Unscheduled DNA synthesis: negative results. Mouse micronucleus test: negative results
Carcinogenicity	No test information

| **Reproductive toxicity and developmental effects (teratogenicity)** | No embryo toxicity or teratogenicity effects detected in rats (EPA 83-3) |

REGULATORY INFORMATION (USA, Canada, Europe)

EU classification and labelling

EU classification	Xi, Irritant
Risk phrases	R36/38 Irritating to skin and eyes. R43 May cause sensitisation by skin contact
Safety phrases	S26 In case of contact with eyes, rinse immediately with plenty of water and seek medical advice. S28 After contact with skin, wash immediately with plenty of soap and water. S36/37/39 Wear suitable protective clothing, gloves and eye/face protection

OSHA hazard communication evaluation (USA)

Meets the criteria for hazardous material as defined by 29 CFR 1910.1200

WHMIS classification (Canada)

Class D – Poisonous and infectious material – Division 2: Materials causing other toxic effects – Sub-Division B: Toxic material

REFERENCES USED IN THE COMPILATION OF THIS DOCUMENT

Various suppliers' MSDS

CHEMICAL PRODUCT NAME

Product name:	**Polyaluminium Hydroxide Chloride (PAC)**
CAS number:	**1327-41-9**
Intended use:	As a coagulant in paper making or in water treatment

AIR CONTAMINATION LIMIT VALUES

UK – OES/TWA: 2 mg m^{-3} as Al (soluble salts) 1998
US – TLV/TWA: 2 mg/m^{-3} as Al (soluble salts) 1998

PHYSICAL AND CHEMICAL PROPERTIES

Form colour and odour	Clear, yellowish liquid. Slight odour
Melting point/boiling point	Boiling point > 100°C
Vapour pressure	Not known
Solubility	Completely soluble in water, although flocculation can occur
pH	< 2 (This cannot be analysed by conventional methods because of its unique composition and high ion-strength. Specialised methods have to be used. When pH is calculated it will be > 2
Viscosity	10–50 mPa.s (20°C)
Partition coefficient (log P_{ow})	Not relevant
Oxidising properties	No oxidising properties

ECOTOXICOLOGICAL TESTS AND ENVIRONMENTAL EFFECTS

Aquatic toxicity
The product's ability to reduce the pH of the surroundings is the reason for the biggest ecological effects to fish, plankton and stationary organisms. At a pH of 3, the mucous membranes of the gills will coagulate, which causes suffocation. A pH of 4 causes irritation to the gills. If the product is not neutralised, prior to release, it can be toxic to aquatic organisms

Biodegradability
Not relevant for an inorganic compound. However, this product will mineralise immediately under normal conditions

Bioaccumulation potential
Unlikely to bioaccumulate

Mobility
No information available

TOXICOLOGICAL TESTS AND POTENTIAL HEALTH EFFECTS

<u>Acute effects</u>

Acute oral toxicity	No test information. However, ingestion would irritate the mucous membranes owing to its low pH
Acute dermal toxicity	No test information
Acute inhalation toxicity	No test information

<u>Irritancy and corrosive effects</u>

Skin irritation/Corrosion	Non-irritating (rabbit, OECD 404). May cause redness in sensitive skin. Repeated or prolonged exposure may defat the skin and cause dermatitis
Eye irritation/Corrosion	No test data available. However, experience from people exposed to the product indicates that it is irritating to the eye. Symptoms include transient redness and pain

Acute respiratory irritation	No test information available. However, aerosols and mists may cause irritation to the respiratory tract
Chemical hypersensitivity	No test information available
Long-term effects	
Subchronic/chronic toxicity and other effects	No information. However, there are no known cases of adverse effects arising from long-term exposure
Mutagenicity	No test information
Carcinogenicity	No test information. Not listed by IARC, NTP or OSHA
Reproductive toxicity and developmental effects (teratogenicity)	No information

REGULATORY INFORMATION (USA, Canada, Europe)

EU classification and labelling

EU classification	Xi, Irritant
Risk phrases	R36 Irritating to eyes
Safety phrases	S26 In case of contact with eyes, rinse immediately with plenty of water and seek medical advice. S39 Wear eye/face protection

OSHA hazard communication evaluation (USA)

Meets the criteria for hazardous material as defined by 29 CFR 1910.1200

WHMIS classification (Canada)

Falls under criteria of Class D – Poisonous and infectious material – Division 2, Sub-Division B. Toxic material

Chemical inventory listings (Y = YES, N = NO)

TSCA	Y	**MOE (Korea)**	Y	**Japan (MITI)**	Y
EINECS	Y	**DSL (Canada)**	Y	**AICS**	Y

REFERENCES USED IN THE COMPILATION OF THIS DOCUMENT

EKA Chemicals AB
Industrial Chemicals Division
Safety Data Sheet 'Ekoflock'
CAS number 1327-41-9 (October 1999)

Association Générale des Hygiénistes et Techniciens Municipaux
TG2 Aluminium salts
Draft Standard for Polyaluminium hydroxide chloride and hydroxide chloride sulphate
CEN/TC 164/WG9 'Water Treatment', Document number 74

CHEMICAL PRODUCT NAME

Product name: **Polyamide Amine Epichlorohydrin Resin (20% aqueous)**
Intended use: Wet strength agent used in the manufacture of paper and board

AIR CONTAMINATION LIMIT VALUES

Epichlorohydrin:
OSHA PEL: 8h TWA 5 ppm (19 mg m^{-3}) skin
ACGIH TLV-TWA: 1.9 mg m^{-3} A3, skin
OEL – UK: TWA 2 ppm (8 mg m^{-3}), STEL 5 ppm, skin
OEL – Sweden: TWA 0.5 ppm (1.9 mg m^{-3}); STEL 1 ppm (4 mg m^{-3}); skin

PHYSICAL AND CHEMICAL PROPERTIES

Form colour and odour	Pale yellowish liquid with a faint odour
Melting point/boiling point	Boiling point > 100°C
Vapour pressure	< 2500 Pa at 20°C
Solubility	Soluble in water
pH	3–4 (commercial product)
Viscosity	20–80 mPa.s (Höppler, 25°C)
Partition coefficient (log P_{ow})	No information
Oxidising properties	No oxidising properties

ECOTOXICOLOGICAL TESTS AND ENVIRONMENTAL EFFECTS

Aquatic toxicity
Acute toxicity to fish: LC$_{50}$ (*Brachydanio rerio*, 96 h) = 13.7 mg l^{-1}, NOEC = 4.7 mg l^{-1} (OECD 203)
Microtox® Test: EC$_{50}$ (15 min) > 10 g l^{-1}
As with other cationic water-soluble polymers, the toxicity to fish is likely to be attributed to adsorption onto the gills and subsequent disturbance of respiratory function

Biodegradability
Not readily biodegradable (OECD 301D)

Bioaccumulation potential
Inherently biodegradable (OECD 302B)

Mobility
The product strongly adsorbs to solid surfaces such as cellulosic fibres and minerals, whereupon it would be slowly biodegraded. This means that the proportion found in the receiving waters is very low

TOXICOLOGICAL TESTS AND POTENTIAL HEALTH EFFECTS

Acute effects

Acute oral toxicity	LD$_{50}$ (rat) > 2000 mg kg^{-1} (OECD Limit Test, 401)
Acute dermal toxicity	No test information available
Acute inhalation toxicity	No test information available

Irritancy and corrosive effects

Skin irritation/Corrosion	Non-irritating to skin (rabbit, OECD 404)
Eye irritation/Corrosion	Mildly irritating to eyes (rabbit, OECD 405)
Acute respiratory irritation	No test information available. However, heated aerosols and mists may cause slight irritation to the respiratory tract
Chemical hypersensitivity	Not a sensitiser (OECD 406)

Long-term effects

Subchronic/chronic toxicity and other effects	No information available
Mutagenicity	Short term *in vitro* bacterial 'Ames test' with *Salmonella typhimurium* gave negative results (OECD 471)
Carcinogenicity	No test information
Reproductive toxicity and developmental effects (teratogenicity)	No test information

REGULATORY INFORMATION (USA, Canada, Europe)

EU classification and labelling

EU classification	Not classified as dangerous
Risk phrases	None
Safety phrases	None

OSHA hazard communication evaluation (USA)

Does not meet the criteria for hazardous material as defined by 29 CFR 1910.1200

WHMIS classification (Canada)

Does not meet the criteria of a Controlled Product as prescribed by the Controlled Products Regulations

Chemical inventory listings (Y = YES, N = NO)

TSCA	**Y**	**MOE (Korea)**	**Y**	**Japan (MITI)**	**Y**
EINECS	**N (polymer)**	**DSL (Canada)**	**Y**	**AICS**	**Y**

REFERENCES USED IN THE COMPILATION OF THIS DOCUMENT

Various suppliers' MSDS

CHEMICAL PRODUCT NAME

Product name:	**Polyamine (50% aqueous)**
CAS number	**42751-79-1**
Intended use:	Charge modification in paper making

AIR CONTAMINATION LIMIT VALUES

Not established

PHYSICAL AND CHEMICAL PROPERTIES

Form colour and odour	Yellow liquid with a slight odour
Melting point/boiling point	Boiling point $c.$ 100°C
Vapour pressure	No information
Solubility	Miscible in water
pH	3–5 (commercial product)
Viscosity	700–1000 cP at 20°C
Partition coefficient (log P_{ow})	No information
Oxidising properties	No oxidising properties

ECOTOXICOLOGICAL TESTS AND ENVIRONMENTAL EFFECTS

Aquatic toxicity
Acute toxicity to fish: LC_{50} (Fathead minnows, 96 h) = 0.27 mg l^{-1} (static conditions)
Acute toxicity to invertebrates: EC_{50} (*Daphnia magna*, 24 h) = 0.3 mg l^{-1}
Toxicity of cationic polymers is highly dependent on the water chemistry. It is believed that tests carried out in pure water overestimates the true environmental impact of this and other cationic polymers

Biodegradability
No test information. The biodegradability of water-soluble cationic polymers is difficult to establish owing to the precipitation of test material, thus leading to a false result

Bioaccumulation potential
Not expected to occur

Mobility
The product is miscible in water with a vapour pressure that is similar to water and it is not readily biodegradable. It is therefore likely to remain in the aqueous compartment

TOXICOLOGICAL TESTS AND POTENTIAL HEALTH EFFECTS

Acute effects

Acute oral toxicity	LD_{50} (rat) > 2000 mg kg^{-1}
Acute dermal toxicity	No test results available
Acute inhalation toxicity	No test results available

Irritancy and corrosive effects

Skin irritation/Corrosion	May cause slight irritation, especially with repeated or prolonged contact
Eye irritation/Corrosion	Irritating to eyes
Acute respiratory irritation	No test results available. However, inhalation of aerosols or mists may cause irritation owing to the low pH
Chemical hypersensitivity	No test results available

Long-term effects

Subchronic/chronic toxicity and other effects	No test information available
Mutagenicity	No test information available
Carcinogenicity	Not listed by NTP, OSHA or ACGIH
Reproductive toxicity and developmental effects (teratogenicity)	No information

REGULATORY INFORMATION (USA, Canada, Europe)

EU classification and labelling

EU classification	Xi, Irritant
Risk phrases	R36 Irritating to eyes
Safety phrases	S26 In case of contact with eyes, rinse immediately with plenty of water and seek medical advice. S39 Wear eye/face protection

OSHA hazard communication evaluation (USA)

Meets the criteria for hazardous material as defined by 29 CFR 1910.1200

WHMIS classification (Canada)

Meets the criteria of a Controlled Product as prescribed by the Controlled Products Regulations. Class D – Poisonous and infectious material – Division 2, Sub-Division B – Toxic material

Chemical inventory listings (Y = YES, N = NO)

TSCA	Y	MOE (Korea)	Y	Japan (MITI)	Y
EINECS	N (polymer)	DSL (Canada)	Y	AICS	Y

REFERENCES USED IN THE COMPILATION OF THIS DOCUMENT

Various suppliers' MSDS

CHEMICAL PRODUCT NAME

Product name: **Polyethyleneimine (modified)**
CAS number   **26913-06-4**
Intended use: Drainage aid

AIR CONTAMINATION LIMIT VALUES

Not established

PHYSICAL AND CHEMICAL PROPERTIES

Form colour and odour	Pale yellow liquid
Melting point/boiling point	Boiling point $c.$ 100°C. Solidification temperature $c.$ –5°C
Vapour pressure	No information
Solubility	Miscible in water and lower alcohols
pH	7.8–8.7 (commercial product)
Viscosity	500–1000 mPa.s at 20°C
Partition coefficient (log P_{ow})	No information
Oxidising properties	No oxidising properties

ECOTOXICOLOGICAL TESTS AND ENVIRONMENTAL EFFECTS

Aquatic toxicity
Acute toxicity to fish: LC_{50} (*Leuciscius idus*, 96 h) = 1–10 mg l^{-1}, LC_{50} (48 h) = 1 mg l^{-1}
Acute toxicity to invertebrates: EC_{50} (*Daphnia magna*, 48 h) = 0.4 mg l^{-1}
Acute toxicity to algae: EC_{50} (*Selenastrum capricornutum*) = 0.25 mg l^{-1}
Acute toxicity to bacteria: EC_{10} (16 h) = 0.93 mg l^{-1} (DIN 38412 Part 8), EC_{50} (16 h) = 3.6 mg l^{-1} (DIN 38412 Part 8)

Biodegradability
OECD 302B/ISO 9888/EEC 88/302: C. DOC reduction: more than 70% eliminated

Bioaccumulation potential
Not expected to occur

Mobility
Elimination by adsorption on to activated sludge

TOXICOLOGICAL TESTS AND POTENTIAL HEALTH EFFECTS

<u>Acute effects</u>

Acute oral toxicity	LD_{50} (rat) > 5000 mg kg^{-1}
Acute dermal toxicity	No test results available
Acute inhalation toxicity	Test using rats produced no mortalities after 8 h exposure in a highly enriched and/or saturated atmosphere at 20°C
<u>Irritancy and corrosive effects</u>	
Skin irritation/Corrosion	Draize test (rabbits) results showed the product not to be irritating
Eye irritation/Corrosion	Draize test (rabbits) results showed the product not to be irritating
Acute respiratory irritation	No test results available
Chemical hypersensitivity	No test results available

Long-term effects

Subchronic/chronic toxicity and other effects	No test results available
Mutagenicity	No test information available
Carcinogenicity	No test information available
Reproductive toxicity and developmental effects (teratogenicity)	No information

REGULATORY INFORMATION (USA, Canada, Europe)

EU classification and labelling

EU classification	Not classified
Risk phrases	None
Safety phrases	None

OSHA hazard communication evaluation (USA)

Does not meet the criteria for hazardous material as defined by 29 CFR 1910.1200

WHMIS classification (Canada)

Does not meets the criteria of a Controlled Product as prescribed by the Controlled Products Regulations

Chemical inventory listings (Y = YES, N = NO)

TSCA	Y	MOE (Korea)	Y	AICS	Y
EINECS	Y	DSL (Canada)	Y		

REFERENCES USED IN THE COMPILATION OF THIS DOCUMENT

STN Easy Regulated Chemical Lists
http://stneasy.FIZ-Karlsruhe.de

STN Easy Material Safety Data Sheets
OHSN OHS35745 MSDS – OHS: 'Polyethyleneimine'
CAS number 26913-06-4

Various suppliers' MSDS

CHEMICAL PRODUCT NAME

Product name:	**Rosin Size Dispersion (30% casein-based)**
CAS number	**Rosin, fumarated (65997-04-8)**
Intended use:	Sizing for the paper and board industry

AIR CONTAMINATION LIMIT VALUES

Not established

PHYSICAL AND CHEMICAL PROPERTIES

Form colour and odour	White liquid
Melting point/boiling point	Boiling point $c.$ 100°C
Vapour pressure	< 2500 Pa at 20°C
Solubility	Miscible in water
pH	6–6.5 (commercial product)
Viscosity	20–40 mPa.s at 20°C
Partition coefficient (log P_{ow})	Not known
Oxidising properties	No oxidising properties

ECOTOXICOLOGICAL TESTS AND ENVIRONMENTAL EFFECTS

Aquatic toxicity
Acute toxicity to fish: LC_{50} (*Brachydanio rerio*, 96 h) = 10.8 mg l^{-1}
Acute toxicity to bacteria: EC10 (Robra test, *Pseudomonas putia*) = 20,700 mg l^{-1}

Biodegradability
Readily biodegradable (OECD 301B)

Bioaccumulation potential
Not established

TOXICOLOGICAL TESTS AND POTENTIAL HEALTH EFFECTS

Acute effects

Acute oral toxicity	Using the OECD Limit Test with the fumarated rosin adduct gave a result of LD_{50} (rat) > 2000 mg kg^{-1}
Acute dermal toxicity	No test information available
Acute inhalation toxicity	No test information available

Irritancy and corrosive effects

Skin irritation/Corrosion	Tests on the fumarated rosin adduct showed it to be severely irritating to the skin (OECD 404)
Eye irritation/Corrosion	Tests on the fumarated rosin adduct showed it to be severely irritating to the eyes (OECD 405)
Acute respiratory irritation	Inhalation of aerosols or mists may irritate the mucous membranes of the naso-respiratory tract
Chemical hypersensitivity	Tests on the dispersion using the Magnusson & Kligman method (OECD 406) gave positive results

Long-term effects

Subchronic/chronic toxicity and other effects	No test information available. However, there have been no cases of long-term adverse effects reported for rosin size products
Mutagenicity	A negative result was obtained in a short-term genotoxic test (Ames test) when using a fortified rosin adduct (OECD 471)
Carcinogenicity	No test information available. Not listed by IARC, OSHA or ACGIH
Reproductive toxicity and developmental effects (teratogenicity)	No information available

REGULATORY INFORMATION (USA, Canada, Europe)

EU classification and labelling

EU classification	Xi, Irritant
Risk phrases	R43 May cause sensitisation by skin contact. R36 Irritating to the eyes
Safety phrases	S26 In case of contact with eyes, rinse immediately with plenty of water and seek medical advice. S28 After contact with skin, wash immediately with plenty of soap and water. S36/37/39 Wear suitable protective clothing, gloves and eye/face protection

OSHA hazard communication evaluation (USA)

Meets the criteria for hazardous material as defined by 29 CFR 1910.1200

WHMIS classification (Canada)

Class D – Poisonous and infectious material
Division 2, Sub-Division B – Toxic material

Chemical inventory listings (Y = YES, N = NO)

TSCA	Y	**MOE (Korea)**	Y	**AICS**	Y
EINECS	Y	**DSL (Canada)**	Y	**Japan (MITI)**	Y

REFERENCES USED IN THE COMPILATION OF THIS DOCUMENT

Various suppliers' MSDS

CHEMICAL PRODUCT NAME

Product name:	**Sodium Chlorate**
CAS number	**7775-09-9**
Intended use:	On-site production of ClO_2 for pulp bleaching

AIR CONTAMINATION LIMIT VALUES

Not established

PHYSICAL AND CHEMICAL PROPERTIES

Form colour and odour	White/colourless crystals or granules
Melting point/boiling point	Melting point = 248°C/478°F. Decomposition occurs at 250–300°C
Vapour pressure	Not applicable
Solubility	Very soluble in water (728 g l^{-1} water at 20°C). Slightly soluble in alcohol
pH	7 (aqueous solution)
Viscosity	Not applicable
Partition coefficient (log P_{ow})	Not applicable
Oxidising properties	Strong oxidiser

ECOTOXICOLOGICAL TESTS AND ENVIRONMENTAL EFFECTS

Aquatic toxicity
Acute toxicity to fish: LC_{50} (*Pimephales promelas*, 96 h) = 1350 mg l^{-1}, LC_{50} (*Oncorhynchus mykiss*, 48 h) = 2750 mg l^{-1}
Acute toxicity to aquatic invertebrates: EC_0 (*Daphnia magna*, 24 h) = 789 mg l^{-1}, EC_{50} = 1093 mg l^{-1}, EC100 = 1362 mg l^{-1}

Biodegradability
Not relevant for an inorganic compound. However, sodium chlorate is rapidly degraded to sodium chloride and oxygen in the soil under anaerobic conditions, and more slowly in aerobic conditions

Bioaccumulation potential
Plants may accumulate chlorate until tissue death occurs (herbicide activity of chlorate). No evidence of accumulation in animals

Mobility
Sodium chlorate will remain dissolved in the aqueous environment. It can be leached out of the soil

TOXICOLOGICAL TESTS AND POTENTIAL HEALTH EFFECTS

Acute effects

Acute oral toxicity	LD_{50} (rat) = 1200 mg kg^{-1}. The lethal dose for humans is approximately 15–30 g. Ingestion may cause nausea, vomiting, abdominal pain and chemical burns in the stomach and intestines
Acute dermal toxicity	LD_{50} (rabbit) > 10,000 mg kg^{-1}. LD_0 (rabbit) > 2000 mg kg^{-1}
Acute inhalation toxicity	LC_{50} (rat, 1 h) > 28 mg l^{-1}. Dust and mist may cause mild irritation. Repeated exposure to chlorates by inhalation may produce toxic effects which gradually appear over weeks

Irritancy and corrosive effects

Skin irritation/Corrosion	Mildly irritating (Draize test, 500 mg over 24 h). Prolonged or repeated skin contact may result in dermatitis
Eye irritation/Corrosion	Mildly irritating (Draize test, 10 mg dose)
Acute respiratory irritation	Inhalation of dust or mist may cause coughing and mild temporary irritation of the nose and throat

Chemical hypersensitivity	Some allergic reactions have been reported in animal studies (guinea pig, Magnusson & Kligman test). Patch testing on humans, however, did not show any effect
Long-term effects	
Subchronic/chronic toxicity and other effects	Subchronic 90-day feeding study in rats gave a NOEL of 100 mg kg^{-1}. The highest dose used, 1000 mg kg^{-1}, resulted in slight anaemia
Mutagenicity	OECD 471, Ames test, produced negative results both with and without the use of a metabolic activator. OECD 482 gave negative results (up to 10 mg l^{-1} concentration). Mammalian cell gene mutation (CHO) assay gave negative results
Carcinogenicity	Not listed by IARC, NTP, OSHA or ACGIH
Reproductive toxicity and developmental effects (teratogenicity)	No information

REGULATORY INFORMATION (USA, Canada, Europe)

EU classification and labelling

EU classification	O, Oxidising, Xn, Harmful
Risk phrases	R9 Explosive when mixed with combustible material. R22 Harmful if swallowed
Safety phrases	S13 Keep away from food, drink and animal feeding stuffs. S17 Keep away from combustible material. S46 If swallowed, seek medical advice immediately and show container or label

OSHA hazard communication evaluation (USA)

Meets the criteria for hazardous material as defined by 29 CFR 1910.1200

WHMIS classification (Canada)

Class C – Oxidizing material
Class D1B – Poisonous and infectious material, immediate and serious effects – Toxic

Chemical inventory listings (Y = YES, N = NO)

TSCA	Y	MOE (Korea)	Y	AICS	Y
EINECS	Y	DSL (Canada)	Y	Japan (MITI)	Y

REFERENCES USED IN THE COMPILATION OF THIS DOCUMENT

CHEMINFO – CCOHS (97-3) August 1997

Record number: 113 'Sodium chlorate'

EKA Chemicals AB, Bohus, Sweden, Bleaching Chemicals Division
'Sodium chlorate' (October 1998)

Richardson, M. L. & Gangolli, S., eds (1994) *The Dictionary of Substances and their Effects*, Vol. 4, Record number S50 'Sodium chlorate'

Chemical Substances CD version 5.1 (1991–99) Arbetarskyddsnämnden 'Sodium chlorate

IUCLID Data Sheet 'Sodium chlorate', ECB Existing Chemicals (23-10-95)

CHEMICAL PRODUCT NAME

Product name: **Sodium Dithionite**
CAS number   **7775-14-6**
Intended use: Mechanical pulp bleaching

AIR CONTAMINATION LIMIT VALUES

Not established

PHYSICAL AND CHEMICAL PROPERTIES

Form colour and odour	White powder with faint sulphurous odour
Melting point/boiling point	Decomposes at $> 130°C$
Vapour pressure	Not applicable
Solubility	No information
pH	No information
Viscosity	No information
Partition coefficient (log P_{ow})	No information
Oxidising properties	None

ECOTOXICOLOGICAL TESTS AND ENVIRONMENTAL EFFECTS

Aquatic toxicity
Acute toxicity to fish: LC_{50} (Golden Orfe, 96 h) = 46 mg l^{-1}
Acute toxicity to invertebrates: EC_{50} (*Daphnia magna*, 48 h) = 989 mg l^{-1}
Acute toxicity to algae: EC_{50} (*Scenedesmus subspicatus*, 72 h) = 120 mg l^{-1}

Biodegradability
Not relevant for an inorganic compound

TOXICOLOGICAL TESTS AND POTENTIAL HEALTH EFFECTS

Acute effects

Acute oral toxicity	LD_{50} (rat) = 2500 mg kg^{-1}. Ingestion may cause nausea, vomiting, abdominal pain, CNS depression, convulsions
Acute dermal toxicity	LD_{50} (rabbit) $> 10,000$ mg kg^{-1}. Not expected to be absorbed
Acute inhalation toxicity	No information

Irritancy and corrosive effects

Skin irritation/Corrosion	Irritates and may burn
Eye irritation/Corrosion	Irritates and may burn
Acute respiratory irritation	Dust may cause severe irritation to the mucous membranes of the respiratory tract. Inhalation of hot vapours or mists produces severe irritation to the respiratory airways

Long-term effects

Subchronic/chronic toxicity and other effects	Prolonged or repeated exposures may cause damage to the lungs
Mutagenicity	No information available
Carcinogenicity	Not listed by IARC, OSHA or ACGIH
Reproductive toxicity and developmental effects (teratogenicity)	No information

REGULATORY INFORMATION (USA, Canada, Europe)

EU classification and labelling

EU classification	Xn, Harmful
Risk phrases	R22 Harmful if swallowed. R7 May cause fire. R31 Contact with acids liberates toxic gases
Safety phrases	S7/8 Keep container tightly closed and dry. S26 In case of contact with the eyes, rinse immediately with plenty of water and seek medical advice. S28 After contact with the skin, wash immediately with plenty of water. S43 In case of fire, use dry powder. Never use water

OSHA hazard communication evaluation (USA)

Meets the criteria for hazardous material as defined by 29 CFR 1910.1200

WHMIS classification (Canada)

Meets the criteria for a Controlled Product as defined by the Controlled Products Regulations

Chemical inventory listings (Y = YES, N = NO)

TSCA	Y	**MOE (Korea)**	Y	**AICS**	Y
EINECS	Y	**DSL (Canada)**	Y	**Japan (MITI)**	Y

REFERENCES USED IN THE COMPILATION OF THIS DOCUMENT

STN Easy. MSDS – CCOHS 2000
AN 1690678 'Sodium Dithionite'

STN Easy. Regulated Chemical Lists
'Sodium Dithionite' CAS number: 7775-14-6
http://stneasy.FIZ.Karlsruhe.de

Chemical Substances CD version 5.1 (1991–99) Arbetarskyddsnämnden 'Sodium Dithionite'

CHEMICAL PRODUCT NAME

Product name:	**Sodium Hydroxide**
CAS number	**1310-73-2**
Intended use:	Diverse applications in the pulp and paper industry

AIR CONTAMINATION LIMIT VALUES

OSHA – PEL: 8 h TWA $2\,mg\,m^{-3}$
OEL – Sweden: TWA $2\,mg\,m^{-3}$
OEL – UK: STEL $2\,mg\,m^{-3}$

PHYSICAL AND CHEMICAL PROPERTIES

Form colour and odour	Clear, colourless, odourless liquid
Melting point/boiling point	Freezing point: 12°C. Boiling point: 145°C
Vapour pressure	< 1 kPa (20°C)
Solubility	Fully soluble in water. Note that this is a strongly exothermic reaction. Soluble in aliphatic alcohols and glycerine
pH	13–14 (0.5% solution)
Viscosity	Similar to water
Partition coefficient (log P_{ow})	Not applicable
Oxidising properties	No oxidising properties

ECOTOXICOLOGICAL TESTS AND ENVIRONMENTAL EFFECTS

Aquatic toxicity
Acute toxicity to fish: LC_{50} (*Gambusia affinis*, 96 h) = $125\,mg\,l^{-1}$
Acute toxicity to aquatic invertebrates: EC_{50} (Daphnia, 48 h) = $40–240\,mg\,l^{-1}$

Biodegradability
Not relevant for an inorganic compound

Bioaccumulation potential
Not likely to accumulate

Mobility
The product is highly soluble in water and has a low vapour pressure. Therefore, it will have a high mobility in the aqueous environment. It is not considered to be persistent in the environment

TOXICOLOGICAL TESTS AND POTENTIAL HEALTH EFFECTS

<u>**Acute effects**</u>	
Acute oral toxicity	LD_{lo} (rabbit) = $500\,mg\,kg^{-1}$. Corrosive. Causes severe burns and pain to the mouth, throat and gastointestinal tract. May cause life-threatening injuries
Acute dermal toxicity	No test information available
Acute inhalation toxicity	Owing to its corrosive nature, inhalation of any respirable aerosols could cause pulmonary oedema (severe, life-threatening lung damage)
<u>**Irritancy and corrosive effects**</u>	
Skin irritation/Corrosion	Four hours' exposure, 5% solution, dermal (rabbit) produced severe necrosis. Corrosive. Can cause severe burns and permanent scarring
Eye irritation/Corrosion	Draize test, 1% solution (rabbit) gave a score of > 5: Severely irritating. Corrosive. The severity of injury increases with the duration of exposure, concentration and speed of penetration into the eye

Acute respiratory irritation	Inhalation of dust or concentrated mist can cause damage to the upper respiratory tract and to lung tissue. The severity of the effects will depend on the extent of exposure
Chemical hypersensitivity	No test information. There is no evidence of it being a skin sensitiser in humans

Long-term effects

Subchronic/chronic toxicity and other effects	No information
Mutagenicity	Short-term *in vitro* bacterial tests suggests non-mutagenic
Carcinogenicity	No conclusive evidence. Not listed by IARC, NTP, OSHA or ACGIH
Reproductive toxicity and developmental effects (teratogenicity)	No information

REGULATORY INFORMATION (USA, Canada, Europe)

EU classification and labelling

EU classification	Corrosive
Risk phrases	R35 Causes severe burns (50% solution)
Safety phrases	S26 In case of contact with the eyes, rinse immediately with plenty of water and seek medical advice. S37/39 Use suitable gloves and eye/face protection. S45 In cases of accident or if you feel unwell, seek medical advice immediately (show label where possible)

OSHA hazard communication evaluation (USA)

Meets the criteria for hazardous material as defined by 29 CFR 1910.1200

WHMIS classification (Canada)

Class E – Corrosive material

Chemical inventory listings (Y = YES, N = NO)

TSCA	Y	MOE (Korea)	Y	AICS	Y
EINECS	Y	DSL (Canada)	Y	Japan (MITI)	Y
Philippines	Y				

REFERENCES USED IN THE COMPILATION OF THIS DOCUMENT

Chem Advisor® Inc.
Record number: 5086, 'Sodium hydroxide'
Issue date: 97-3 (July 1997)

STN Easy. MSDS – CCOHS 1999
AN 2027031 number 54038

CHEMINFO – CCOHS (97-3) August 1997
Record number: 5 'Sodium hydroxide'

EKA Chemicals AB, Bohus, Sweden, Bleaching Chemicals Division
'Sodium hydroxide 50%'

Richardson, M.L. & Gangolli, S., eds (1994) *The Dictionary of Substances and their Effects*, Vol.4, Record number S68

Chemical Substances CD version 5.1 (1991–99) Arbetarskyddsnämnden 'Sodium Hydroxide'

IUCLID Data Sheet 'Sodium hydroxide', ECB Existing Chemicals 23-10-95

CHEMICAL PRODUCT NAME

Product name:	**Sodium Silicate (37% solution)**
CAS number	**1344-09-8 (molar ratio 3.35)**
Intended use:	Stabiliser for peroxide bleaching of pulp

AIR CONTAMINATION LIMIT VALUES

Not established

PHYSICAL AND CHEMICAL PROPERTIES

Form colour and odour	Colourless, turbid liquid with a slight odour
Melting point/boiling point	Boiling point = $102°C/216°F$. Freezing point $c. -3°C$
Vapour pressure	Approximately 2.2 Pa (20°C)
Solubility	Completely soluble in water
pH	11.3
Viscosity	No information
Partition coefficient (log P_{ow})	Not applicable
Oxidising properties	None

ECOTOXICOLOGICAL TESTS AND ENVIRONMENTAL EFFECTS

Aquatic toxicity

Acute toxicity to fish: LC_{50} (*Lepomis macrochirus*, 96 h) = $301–478$ mg l^{-1}

Acute toxicity to aquatic invertebrates: EC_{50} (*Daphnia magna*, 96h) = 216 mg l^{-1}, $EC100 = 247$ mg l^{-1} (pH 9.1)

Acute toxicity to bacteria: EC_0 (*Pseudomonas putia*, 30 min) > 1000 mg l^{-1} (DIN 38415, Teil 27, corresponds to OECD 209)

General comment: solutions of sodium silicate, which are unbuffered, produce alkaline conditions that can be harmful to aquatic life

Biodegradability

Not relevant for an inorganic compound. However the silicate in the product will mineralise and precipitate at a pH > 9. If this product is not neutralised it can be toxic for aquatic organisms owing to its high pH

Bioaccumulation potential

No potential to accumulate

Mobility

Soluble silicates are ubiquitous in the environment

TOXICOLOGICAL TESTS AND POTENTIAL HEALTH EFFECTS

<u>**Acute effects**</u>

Acute oral toxicity	LD_{50} (rat) > 2000 mg kg^{-1}. Acute oral toxicity depends on the molar ratio of SiO_2 to Na_2O. Ingestion may cause irritation to the mucous membranes of the mouth, oesophagus and stomach
Acute dermal toxicity	No test data
Acute inhalation toxicity	No test data

<u>**Irritancy and corrosive effects**</u>

Skin irritation/Corrosion	Non-irritating (OECD 404) when the sample has a molar ratio of 3.45 and concentration of 35%. However, the resultant effects are based on the molar ratio used and concentration of the product

Eye irritation/Corrosion	Non-irritating (Draize test) when the sample has a molar ratio of 3.2 and concentration of 36%. However, the resultant effects are based on the molar ratio used and concentration of the product. A concentrated solution of molar ratios less than or equal to 2.9 are severely irritating
Acute respiratory irritation	Inhalation of mist or spray may cause coughing and other symptoms of upper respiratory tract irritation
Chemical hypersensitivity	No test results. However, sodium silicate is not expected to be a sensitiser
Long-term effects	
Subchronic/chronic toxicity and other effects	The administration of sodium silicate in the drinking water of Wistar rats over 2 years gave a NOEL of 792 mg kg^{-1}. No specific changes were seen at the end of the 2-year period
Mutagenicity	DNA damage and repair assay using *Bacillus subtilis* recombination. Repair-deficient and wild-type strains without metabolic activation gave negative results. *E. coli* in a reverse mutation assay gave negative results
Carcinogenicity	Not listed by IARC, NTP, OSHA or ACGIH. A 2-year study in rats administered sodium silicate in drinking water gave no carcinogenic effects
Reproductive toxicity and developmental effects (teratogenicity)	No conclusive tests

REGULATORY INFORMATION (USA, Canada, Europe)

EU classification and labelling

EU classification	No classification requirements for solutions with molar ratios > 3.2 and concentrations < 40%
Risk phrases	None
Safety phrases	None

OSHA hazard communication evaluation (USA)

Does not meet the criteria for a hazardous chemical as per OSHA Hazard Communication Standard 29 CFR 1910.1200

WHMIS classification (Canada)

Not a controlled product

Chemical inventory listings (Y = YES, N = NO)

TSCA	Y	MOE (Korea)	Y	AICS	Y
EINECS	Y	DSL (Canada)	Y	Japan (MITI)	Y

REFERENCES USED IN THE COMPILATION OF THIS DOCUMENT

Various suppliers' MSDS

STN Easy. Regulated Chemical Lists
'Sodium Silicate' CAS number: 1344-09-8
http://stneasy.FIZ.Karlsruhe.de

IUCLID Data Sheet 'Silicic Acid, Sodium Salt, CAS number: 1344-09-8
ECB Existing Chemicals 23-10-95

CHEMICAL PRODUCT NAME

Product name: **Stearic Acid**
CAS number   **57-11-4**
Intended use: De-inking aid

AIR CONTAMINATION LIMIT VALUES

Not established

PHYSICAL AND CHEMICAL PROPERTIES

Form colour and odour	White/slightly yellow crystals or powder with slight tallow odour
Melting point/boiling point	Boiling point 376°C/709°F. Melting point 69°C/157°F
Vapour pressure	Very low at 25°C; 1 mmHg at 173°C
Solubility	Practically insoluble in water. Very soluble in ether, acetone, ethanol and toluene
pH	11.3 (commercial product)
Viscosity	< 25 mPa.s (20°C)
Partition coefficient (log P_{ow})	No information
Oxidising properties	No oxidising properties

ECOTOXICOLOGICAL TESTS AND ENVIRONMENTAL EFFECTS

Aquatic toxicity
Acute toxicity to fish: LC_{50} (guppy) = 14 mg l^{-1}

Biodegradability
BOD_5 = 1.44 mg O_2 l^{-1}

Bioaccumulation potential
Incorporated into animal fat

TOXICOLOGICAL TESTS AND POTENTIAL HEALTH EFFECTS

Acute effects

Acute oral toxicity	LD_{50} (rat) > 5 g kg^{-1}. Ingestion of very high levels may cause nausea and a laxative effect
Acute dermal toxicity	LD_{50} (rabbit) > 5 g kg^{-1}
Acute inhalation toxicity	No test information. However, will probably cause little adverse effect in the lungs or systemic effects

Irritancy and corrosive effects

Skin irritation/Corrosion	0.5 g caused no irritation in the standard Draize test. However, repeated or prolonged contact with dust or concentrated solution may cause dermatitis
Eye irritation/Corrosion	In tests, application of 0.1 mg did not produce irritation to the eyes. However, dust or mist may be slightly irritating to the eyes
Acute respiratory irritation	High concentration of dust may cause coughing and temporary irritation to the respiratory tract
Chemical hypersensitivity	7% stearic acid in petrolatum applied to the skin of 26 human volunteers produced no allergic reactions

Long-term effects

Subchronic/chronic toxicity and other effects	Stearic acid is the normal component of many foods. Animal studies where large levels of stearic acid were administered produced only slight effects
Mutagenicity	Short-term *in vitro* bacterial 'Ames test' gave negative results, both with and without metabolic activation. No human or animal *in vivo* tests available
Carcinogenicity	Two carcinogenic animal studies showed stearic acid not to be tumorigenic. Not listed by OSHA, IARC, NTP
Reproductive toxicity and developmental effects (teratogenicity)	No test information. However, not expected to cause reproductive embryotoxic or teratogenic effects

REGULATORY INFORMATION (USA, Canada, Europe)

EU classification and labelling

EU classification	Not classified as dangerous
Risk phrases	None
Safety phrases	None

OSHA hazard communication evaluation (USA)

Does not meet the criteria for a hazardous material as defined by 29 CFR 1910.1200

WHMIS classification (Canada)

Does not meet the criteria of a Controlled Product as prescribed by the Controlled Products Regulations

Chemical inventory listings (Y = YES, N = NO)

TSCA	Y	MOE (Korea)	Y	AICS	Y
EINECS	Y	DSL (Canada)	Y		

REFERENCES USED IN THE COMPILATION OF THIS DOCUMENT

Richardson, M. L. & Gangolli, S., eds (1994) *The Dictionary of Substances and their Effects*, Vol. 4, Record number S108 'Stearic Acid'

Chemical Substances CD version 5.1 (1991–99) Arbetarskyddsnämnden 'Stearic Acid'

CHEMINFO – CCOHS (97-3) August 1997
Record number: 544 'Stearic Acid'

STN Easy. Regulated Chemical Lists
'Stearic Acid' CAS number: 57-11-4
http://stneasy.FIZ.Karlsruhe.de

CHEMICAL PRODUCT NAME

Product name:	**Styrene–Acrylate Copolymer (aqueous dispersion)**
Intended use:	Synthetic surface sizing for paper industry

AIR CONTAMINATION LIMIT VALUES

Not established

PHYSICAL AND CHEMICAL PROPERTIES

Form colour and odour	Light brown liquid with the characteristic odour of styrene
Melting point/boiling point	Boiling point *c.* 100°C
Vapour pressure	Similar to water
Solubility	Miscible in water
pH	3–5 (commercial product)
Viscosity	< 100 cP
Partition coefficient (log P_{ow})	No information
Oxidising properties	No oxidising properties

ECOTOXICOLOGICAL TESTS AND ENVIRONMENTAL EFFECTS

Aquatic toxicity
Acute toxicity to fish: LC_{50} (*Leuciscius idus*, 96 h) > 1000 mg l^{-1} (OECD 203)
Acute toxicity to bacteria: EC_{10} = 560 mg l^{-1}, EC_{50} = 2600 mg l^{-1}

Biodegradability
Not readily biodegradable (21% after 28 days, OECD 301B)

Bioaccumulation potential
Not expected to occur

Mobility and fate
The product is miscible in water with a vapour pressure which is similar to water and it is not readily biodegradable. It is therefore likely to remain in the aqueous compartment

TOXICOLOGICAL TESTS AND POTENTIAL HEALTH EFFECTS

Acute effects

Acute oral toxicity	LD_{50} (rat) > 2000 mg kg^{-1}
Acute dermal toxicity	No test results available
Acute inhalation toxicity	No test results available

Irritancy and corrosive effects

Skin irritation/Corrosion	No test results available. However, may cause mild, but transient, irritation to the skin
Eye irritation/Corrosion	No test results available. May cause mild, but transient, irritation to the eyes
Acute respiratory irritation	No test results available. However, inhalation of aerosols or mists may cause irritation owing to the low pH
Chemical hypersensitivity	No test results available

Long-term effects

Subchronic/chronic toxicity and other effects	No test information available
Mutagenicity	No test information available
Carcinogenicity	No test information available
Reproductive toxicity and developmental effects (teratogenicity)	No information

REGULATORY INFORMATION (USA, Canada, Europe)

EU classification and labelling

EU classification	Not classified
Risk phrases	None
Safety phrases	None

OSHA hazard communication evaluation (USA)

Does not meet the criteria for a hazardous material as defined by 29 CFR 1910.1200

WHMIS classification (Canada)

Does not meet the criteria of a Controlled Product as prescribed by the Controlled Products Regulations

Chemical inventory listings (Y = YES, N = NO)

TSCA	Y	**MOE (Korea)**	N	**AICS**	N
EINECS	Y	**DSL (Canada)**	N	**Japan (MITI)**	Y

REFERENCES USED IN THE COMPILATION OF THIS DOCUMENT

Various suppliers' MSDS

7 Glossary of terms

In the list given below these are a number of terms commonly encountered in the health and safety literature. These terms do not comprise an exhaustive list – such a glossary would probably run into hundreds of pages! Therefore it is suggested that this particular glossary is used as a starting point allowing you to create your own list tailored to your own work situation.

Absorption	The uptake of a chemical or water into or across a tissue, such as skin.
Accumulation	This is the build-up of chemical in the organism as a result of repeated or long-term exposure.
ACGIH	American Conference of Governmental Industrial Hygienists which establishes exposure limits for workers (Threshold Limit Values, TLVs).
Acute toxicity	Adverse effects arising from a single exposure or short-term exposure to a chemical.
ADI	Acceptable Daily Intake. This is the maximum amount of chemical anticipated to cause adverse health effects in humans when daily ingested over the whole lifetime.
Adjuvant	A substance which is used to enhance or modify the immune response in a non-specific manner. Freund's Complete Adjuvant is such an example.
Adsorption	The process by which chemicals become attached to solid surfaces.
Aerobic	With oxygen.
Aetiology	The study of the causes of disease.
AICS	Australian Inventory of Chemical Substances (Australian equivalent to EINECS).
Allergen	A substance causing an allergic reaction.
Allergy	An adverse reaction which is caused by an over-stimulation of the immune system in response to a specific allergen which is otherwise harmless and would normally be tolerated by the majority of those who come in contact with it.
Ames Test	A short-term *in vitro* test for genotoxicity.
Anaerobic	Conditions which are without oxygen.
Antigen	A substance that elicits a specific immune response.
Aqueous	Pertaining to water.
Benign tumour	A tumour that does not invade surrounding tissues or spread to other parts of the body, i.e. non-cancerous.

Bioaccumulation The process by which a chemical is accumulated in an organism.

Bioconcentration The process by which chemicals enter the tissues of organisms from the environment and are accumulated to levels greater than the medium in which the organism resides.

Biodegradation The breakdown of the chemical into simpler compounds by micro-organisms. Does not apply to inorganic compounds.

Biomagnification The process by which a chemical moves up the food chain and at each step (trophic level) becomes more concentrated than the previous.

Black list This is an EC list of priority chemicals that are likely candidates for control owing to both the volume of production and also the hazard to the aquatic environment. It is List 1 annexed to EEC Directive 76/464/EEC. Pollution by 'black list' chemicals should be eliminated.

BOD Biological Oxygen Demand. This measures the amount of oxygen that is used by micro-organisms as they break down organic material. It provides an indication of potential biodegradation.

Carcinogen A chemical capable of inducing cancer.

Carcinogenesis The process of the development of cancer.

Cardiovascular system Often called the 'CV system'. This includes the heart and blood vessels.

CAS Number Abbreviation for Chemical Abstracts Service Registry Number. This is a universal number that is assigned to a specific compound. It is used to overcome any problems in nomenclature or language.

Cell The functional unit of an organism. Comprises a nucleus which is surrounded by a nuclear membrane and cytoplasm.

Central nervous system Often abbreviated to 'CNS', it includes the brain and spinal cord.

Chromosome Material contained within the cells comprising genetic information. Its main constituent is DNA.

Chronic toxicity Adverse health effects arising from continuous or intermittent exposure to low concentrations of chemical over a lifetime.

Clastogen A chemical causing breaks in chromosomes.

COD Chemical Oxygen Demand.

Code of Federal Regulations The entire compilation of USA regulations. For example, 29 CFR 1910.1200 is the Hazard Communication Standard and 21 CFR 100 to 199 is the section containing most of the regulations pertaining to food and food contact materials, e.g. paper.

Congenital Disease or defect that exists from birth.

Conjunctivitis Inflammation of the conjunctiva, which is a membrane that lines the inner surface of the eyelid.

Control group	This is a group of experimental organisms that are not exposed to the treatment or chemical used in the main test. They are compared to the exposed experimental groups in order to observe whether the resulting effects in the experimental groups are significant.
Corrosive	A chemical that causes irreversible alterations or destruction in living tissue at the site of contact.
Cumulative exposure	A summation of all the exposures that have been undergone by an organism during a specified period of time.
Degradable	Can be broken down into simpler structures.
Dermatitis	Inflammation of the skin.
Dermatitis	An inflammatory response of the skin.
Developmental toxicology	Adverse toxic effects in the developing embryo or foetus.
DOC	Dissolved Organic Carbon.
Dose	The amount of chemical administered. It is a measure of exposure and is usually expressed as milligrams per kilogram of test organism or as parts per million (ppm).
Dose–Response	The relationship between the dose of a chemical and the degree/severity of the resulting effect.
DSL	Domestic Substances List. Canadian Inventory of Chemical Substances (equivalent to EINECS).
EC$_{50}$	Median Effective Concentration, 50%. This is the concentration at which a biological effect is seen in 50% of the test group.
Ecotoxicology	The study of the adverse effect of chemicals on the environment (or more precisely, ecosystems).
EINECS	European Inventory of Existing Commercial Chemical Substances. All existing substances on the European market in the 10-year period to September 1981 were put on an inventory. It excluded polymers and contains approximately 100,000 substances, all called by their proper chemical name together with the CAS and EINECS numbers.
Elimination/excretion	The removal of a chemical from the body. This occurs mostly via exhalation of air, or in the urine or faeces.
Embryo	The name given to the developing young during the early stages of intrauterine development.
End-point	A specific biological effect or response which is used as an indicator of the effect of a chemical on the organism. For example, lethality can be the end-point, or a change in enzyme function, etc.
Environmental fate	The final destination of a chemical once it has been released into the environment. Aspects such as transport in the different environmental compartments have to be taken into consideration as well as bioaccumulation potential and persistence.

Epidemiology   The study of the incidence and distribution of disease in populations.

Etiology   The study of the cause of disease.

Eukaryotic cell   A cell in which the genetic material is contained within a distinct membrane.

Exposure   Contact with a chemical. In the industrial setting, the three most common routes are inhalation, ingestion (orally) and by skin (dermal) contact.

Extrapolation   The process of estimating unknown values by the calculation from known values.

Foetus   The developing young usually from the first trimester (in humans) after conception until birth.

Gametes   Mature germ cells. These are capable of creating a new individual during sexual reproduction.

Gavage   Forced feeding by use of a tube.

Gene   The basic unit of heredity.

Genotoxic   Chemicals that cause hereditary mutations as a result of directly interacting and altering or modifying the structure of the DNA.

Genotoxic carcinogens Chemicals causing cancer as a result of directly interacting with DNA.

Germ cells   A cell containing only half the number of chromosomes (haploid cell). These cells of the reproductive system can give rise to a new organism. They are the sperm cells and ovum in humans.

Gestation   The duration of pregnancy.

Gonad   A reproductive organ of animals that produces gametes.

Grey list   List 2 annexed to EEC Directive 76/464/EEC that includes compounds such as certain biocides, heavy metals, cyanides and fluorides. Pollution by these compounds must be reduced.

Haploid   A cell containing only one set of chromosomes. A diploid cell contains two sets, one from each parent.

Hazard   The intrinsic nature of a chemical to cause an adverse effect.

Haemotoxic   A chemical causing toxic effects to one or more of the components of blood or a change in the properties of the blood such as pH, etc.

Histidine   An amino acid from which histamine is made.

HMIS   Hazardous Materials Identification System.

Hydrolysis   A process by which a chemical reacts with water resulting in its decomposition.

Hypersensitivity   The state in which an individual reacts following exposure to a substance (an allergen) with allergic effects after having been previously exposed (sensitised).

IARC   The International Agency for Research on Cancer.

IgE antibody   A specific type of immunoglobulin which is secreted by B-cells as part of an immune response.

Immune response	The general response of the body to substances that are either foreign or perceived to be so. It can include the production of antibodies, cell-mediated immunity, etc.
In vitro	Latin for 'in glass', these are studies conducted in order to investigate the effects of chemicals by using tissue, cellular or sub-cellular extracts from a living organism.
In vivo	Studies carried out in living organisms.
Intrinsic properties	Inherent to the individual, chemical, etc.
Invertebrate	An organism that does not have a backbone (spinal column).
Irritant	A chemical causing a localised inflammatory reaction to skin or mucous membranes at the site of contact.
IUPAC	International Union of Pure and Applied Chemistry (among other things responsible for setting rules on chemical nomenclature).
KECL	Korean Existing Chemicals Inventory (equivalent to EINECS).
Latent period	This is the period of time from which exposure first occurs to the appearance of an adverse effect.
LC$_{50}$	Lethal Concentration, 50%. This can be used for describing acute fish toxicity or acute inhalation toxicity. It is the concentration which causes death in 50% of the test population.
LD$_{50}$	Lethal Dose, 50%. This is the dose causing death in 50% of the test population.
Lethality	Death.
Lipophilic	'Fat-loving'.
LOEC	Lowest Observable Effect Concentration. The lowest observable concentration at which an effect is observed.
LOEL	Lowest Observable Effect Level.
Lymphocyte	A type of white blood cell involved in the immune response.
MAC value	Maximale Aanvaarde Concentratie, or maximum permissible concentration, for the Netherlands. Usually expressed as a time-weighted average over eight hours. It is legally binding.
MAK value	Maximale Arbeitsplatz Konzentrationen (the German maximum concentration value in the workplace).
Malignant tumour	A tumour which is cancerous.
Meiosis	Cell division in reproductive cells (sperm or egg).
Metabolism	The production of energy as a result of various chemical processes in the body. Chemicals can be metabolised once they are absorbed into the body resulting in structural changes and the production of metabolites.
Mitosis	Cell division in non-reproductive cells.
MSDS	Material Safety Data Sheet. Also known as a Safety Data Sheet.

Mutation	An alteration in the genetic material, which can be passed on to subsequent generations.
Necropsy	A term used instead of autopsy when referring to animals.
Neoplasm	Another word for tumour.
Neurotoxin	A chemical that causes adverse effects to the nervous system.
NFPA	The National Fire Protection Association (NFPA). The NFPA 704 hazard rating system is used for indicating the health, reactivity and flammability hazards of chemicals along with a precautionary symbol used where necessary. This information is presented pictorially within the form of a diamond.
NIOSH	National Institute of Occupational Safety and Health.
NOAEC	No Observable Adverse Effect Concentration. This is the largest concentration not producing any adverse effects.
NOAEL	No Observable Adverse Effect Level. This is the largest dose that does not produce any adverse effects.
Non-genotoxic carcinogens	Chemical carcinogens that cause cancer by effects other than direct damage to the genetic material.
NTP	National Toxicology Programme.
OECD	Organisation for Economic Co-operation and Development.
OES	Occupational Exposure Standard
Oral	By mouth.
Organogenesis	The time period during embryo development when all the major organ systems and organs are formed. It is at this time that the embryo is most susceptible to anything interfering with this development.
OSHA	Occupational Safety and Health Administration (USA).
Ovum	Female reproductive (egg) cell.
PEL	Permissible Exposure Limit. OSHA air contamination standard.
Perinatal	The time immediately before and after birth.
PICCS	Philippines Inventory of Controlled Chemical Substances (equivalent to EINECS).
Postnatal	The time period after birth.
Preparation	An intentional physical mixture of two or more substances between which no chemical reaction takes place.
Primary irritant	A chemical that causes an inflammatory reaction on first contact.
Prokaryotic cell	A cell where the genetic material is not separated from the cytoplasm by a nuclear membrane.
Red list	This is a list of substances (23 in all) specific to the UK, the release of which is proscribed for Integrated Pollution Control.
Respiration	The oxidation of organic materials in living organisms resulting in the release of energy and carbon dioxide.

Respiratory tract	Includes the nose, trachea, larynx and lungs, essentially all the passages through which air is inhaled and exhaled during breathing.
Rhinitis	Inflammation of the nasal mucous membranes.
Risk	The likelihood that an adverse effect will occur.
SAR	Abbreviation for Structure–Activity Relationships.
SDS	Safety Data Sheet. Also known as Material Safety Data Sheet.
Sensitisation	The immune process by which individuals become hypersensitive to a substance to which they are exposed. Subsequent exposures can lead to the development of an allergy.
SEPA	The inventory of existing chemical substances in China, drawn up in 1999 for substances manufactured or used in China between 1 January 1992 and 31 December 1996.
Somatic cells	Non-reproductive cells of the body. These contain the full complement of chromosomes (diploid cells).
STEL	Threshold Limit Value–Short Term Exposure Limit. The ACGIH defines this as the time-weighted average (TWA) airborne concentration to which a worker may be exposed for periods of up to 15 minutes. However, there should be no more than four excursions per day and there should be at least 60 minutes between them.
Subcutaneous	Under the skin.
Subchronic/subacute toxicity	The adverse effects arising from either repeated or daily dosing or exposure to a chemical for part of the life-span (usually not more than 10% of the life-span).
Substance	Chemical elements and their compounds as they occur in the natural state or as produced by industry, including any additives required for the purpose of placing them on the market (EINECS).
Systemic toxicity	Pertains to affecting the body as a whole.
Target organ effects	Chemically induced adverse effects on specific organs or systems of the body. For example toluene is a hepatotoxin (causes adverse effects on the liver).
TD$_{lo}$	The lowest dose that produces signs of toxicity in humans, including tumours or reproductive effects in humans or animals when introduced by any exposure route except inhalation (for inhalation TC$_{lo}$ applies).
Teratogen	A chemical agent causing adverse effects in the normal embryonic development without causing lethality in the foetus or maternal toxicity.
ThOD	Theoretical Oxygen Demand. This is the theoretical amount of oxygen used to completely mineralise a test compound.
Threshold	The dose of a chemical above which a measurable observed effect is seen and below which none is seen.

TLm TL$_{50}$	The median Tolerance Limit. TLm is usually expressed in ppm. It designates the concentration at which the test organisms survive. For example, TL$_{50}$ is the concentration at which 50% of the test population survive.
TLV	Threshold Limit Value – a term used by the ACGIH to describe the maximum airborne concentration of a material to which most workers can be exposed during their normal work without adverse health effects.
TLV-TWA	Threshold Limit Value – Time-Weighted Average, the allowable time-weighted average concentration for a normal 8-hour day or 40-hour week.
Toxicity	The capacity to cause injury on a living organism, the severity of which depends on the exposure concentration and duration.
Toxicology	The study of the adverse effects of chemicals on living organisms.
TSCA	Toxic Substances Control Act. US law governing dangerous (hazardous) substances (equivalent to EINECS), effective from 1 January 1977. Controls the exposure to and use of raw industrial chemicals not subject to other laws. New chemicals are to be evaluated prior to manufacture and can be controlled based on risk.
TWA	Time-Weighted Average.

Index